KIN FOLK

Charleston, SC
www.PalmettoPublishing.com

Kin Folk
Copyright © 2023 by Jim Thorpe IV

Published by: Demery Publishing, LLC
Post Office Box 335
Vine Grove, KY 40175
(502) 377-1458 http://www.demernassociates.com

Cover Design by: Info@sprintart.com Printed in the United States of America

All rights reserved

No portion of this book may be reproduced, stored in a retrieval system, or transmitted in any form by any means–electronic, mechanical, photocopy, recording, or other–except for brief quotations in printed reviews, without prior permission of the author.

Paperback ISBN: 978-0-9819411-2-7

KIN FOLK

JIM THORPE IV

Staff Sergeant (SSG) James Dayshawn Hadley
(otherwise known as Hat),
Richard Hoehne,
and Sergeant (Sgt.) Martin Middleton
have been friends since the first grade.

This story is about them.

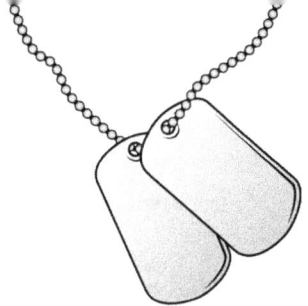

CHAPTER ONE

When SSG Hadley ("Hat") and Sgt. Middleton arrived in Iraq for the third time, things seemed different. The dust, heat, sun, and wind were the same, but the battle had changed.

The first firefight of that tour demonstrated the insurgents' new tactics. One night after a patrol, Hat returned to our tent after being summoned to the Tactical Operation Center (TOC). He informed us combat outpost #7 was under heavy rocket attack, and the sandstorm and high winds were preventing airborne support or artillery. As the quick reaction platoon, we were called to action, and two squads were to engage - our squad and that of Staff Sergeant Lewis, who was new to our platoon. After Hat told us our mission, Sgt. Middleton started the pre-combat check rundown, while Hat sat down with SSG Lewis to go over our battle plan.

After completing our pre-combat check rundown, Sgt. Middleton told Private First-Class Brad Frost to ensure Hat's vehicle fuel was topped off and the two eight-gallon fuel cans were filled to the brim. Private First-Class Brad Frost? That's me; the good-looking redneck from Minnesota and the proud driver of the up-armored United States

Military High Mobility, Multi-purpose wheeled vehicle (HMMWV) - pronounced "HUMVEE."

I was the driver for Hat, and our HUMVEE was numbered with a three. As the first squad leader, Hat's HUMVEE was supposed to be numbered with a one but, I asked Hat if we could change our number to three for Dale Earnhardt. To my surprise, Hat said his favorite driver was the Intimidator and Jeff Gordon. I was surprised by his reply and happy he appreciated the number three. HUMVEE number three had extra steel plates, making it heavier and more fuel deficient than other HUMVEE's. SGT Middleton had told me that after completion of their second tour in Iraq, Hat asked the guys in the motor pool to add re-enforced steel around the wheel wells, top, and sides of the HUMVEE. He also asked them to remove all plastic, add steel stripes in the windows, and replace the tires with a new airless tire he saw on the internet. The guys in the motor pool had that HUMVEE looking like the bus from the Dirty Harry Movie! When Hat picked up the HUMVEE two weeks later, the HUMVEE front view had six 10 inches by 5 inches wide steel stripes for the driver and front passenger. The sides had two of those steel stripes and the back had three of the same. The motor pool had also added heavy duty shocks, raised the HUMVEE, and added a trap door in the floor. That was one special vehicle…

"Private Frost!" Sgt. Middleton screamed, "Come with me to the fuel point!" And he again re-enforced the message of always checking the fuel. I don't like being yelled at, but this was a combat zone, and I knew that politeness was kind of thrown out the window. Besides, my dream was to someday drive for Joe Gibbs racing, watching the fuel was good advice...

As we departed for outpost number 7, I heard Hat tell SSG Lewis "When we come to the fork in the road, go one mile north, and then turn east towards outpost number 7. My squad will be heading down the one lane road to outpost number 7."

Hat knew we were going to get hit – hell, we all knew it, but all the veterans were calm; Hat, Sgt's Middleton, Angel Burgos ("AB"), Joey Pisciotta ("Joey P") and Benjamin Saint James. Those guys had been together ever since basic training. Utah and Rick Hoehne were also in basic training with them but, now they were back in the States after their 2nd tour in Iraq. The Vets always took care of the newbies, and Hat would always choose to take with him any new soldiers he thought might freeze up in combat. It's just the way he worked.

As we entered a street that I knew became narrow then widened out at the end, Hat radioed "Guinea Pig!" on our ear-mouth wireless

microphones. We heard "Roger," from Sgt. Middleton in HUMVEE number 2, and "Roger," from AB in HUMVEE number 1. Hat insisted I drive slowly down the street, and Sgt. Middleton's HUMVEE followed even slower, allowing us about 75 feet of distance. Hat knew from experience the insurgents liked to use Rocket Propelled Grenades at close range, and the "guinea pig" routine was one we knew well.

By sending in one vehicle, and giving that first vehicle room, it allowed Sgt. Middleton's HUMVEE to return initial kill fire without being fired upon. Hat counted on Sgt. Middleton's HUMVEE to fire lethal rapid return fire, giving the "guinea pig" HUMVEE time to return fire.

Meanwhile, AB and Benjamin stopped their HUMVEE's at the beginning of the street. AB posted Specialist Clayton Barnett to the left front of the HUMVEE, and Private First-Class David Cousins to the right rear. Benjamin told them to shoot anything moving from their directions, but not to shoot down the street. Benjamin and AB both failed sniper school during week five of a seven-week course. No one was sure how the hell that happened because both Benjamin and AB were deadly shooters; one shot, one kill. Benjamin and AB removed the roped grappling hooks and threw the hooks onto the roof on each side of the building and climbed the rope to the roof. The last HUMVEE covered our back door or as we say our 6.

My HUMVEE was about 35 yards from the end of the street. I hooked our speakers on the outside of our HUMVEE, and Hat hit the switch on the MP3 player. Hat previously asked an interpreter to rap in Arabic over a Timbaland hyped Middle Eastern beat. The words of the rap were repeated.

"You can't hide from me, you can't hide from me, you can't hide from me, I'm gonna get ya, I'm gonna get ya, you can't hide from me, you can't hide from me, you can't hide from me, I'm gonna get ya, I'm

gonna get ya – so just show yourself, be yourself, it's alright, ya gonna die tonight cause, you can hide from me, you can't hide from me, you can't hide from me, I'm gonna get ya, I'm gonna get ya, if you believe the shit you always spit, the hatred, jealousy and envy, then show yourself and be yourself cause, you can't hide from me, you can't hide from me, you can't hide from me, I'm gonna get ya, I'm gonna get ya" over and over.

Of course, the music made the insurgents direct all their attention onto our HUMVEE and not on Sgt. Middleton's. When our HUMVEE was about 30 yards from the end of the street, windows in the homes opened, and gunmen armed with RPG's and AK47's unloaded on our HUMVEE.

The noise was deafening as the RPG's and AK-47's rocked our HUMVEE, but the insurgents were in for a big surprise. Sure enough, soon Sgt. Middleton and his group opened on them with the M60 machine gun and everything else they had. Their response gave us time to crawl out of our HUMVEE and blaze away with the 50-caliber machine gun. Sgt. Middleton then drove down the street, as those located in the homes in the middle started firing on our position. Joey Pisciotta was Sgt. Middleton's gunner, and when Joey fired on those middle section homes with the 50-caliber machine gun, WHAT! With the bullets flying and the bodies started to fall out of the windows and doorways it was like red snowflakes.

We backed-up past Sgt. Middleton's position and repositioned ourselves at a location that would enable us to return fire at the windows at the beginning of the street, and just like clockwork, the insurgents opened their windows and soon Hat's gunner SPC Lonny Johnson was yelling "Donald Gordon, Motherfucker!" while shooting his M240 machine gun.

The insurgents' new tactic included a feeble attempt to come from behind to attack us. Hat knew their tactics, however, and he used their own tactics against them. AB and Benjamin positioned on the roof were picking them off one at a time: If it moved, they shot it. No one got hit, but PFC Royster said his ears were ringing because our HUMVEE took a lot of hits! He was the communications specialist and the greenest soldier, but even with ringing ears he did well opening-up on the insurgents with his M4. They had a three-pronged attack, but soon discovered we had an ass-kicking answer for each tactic.

After a fire fight, it was normal for us to search the bodies and gather intelligence but, our mission was helping outpost number 7, so Hat called and said, "Mount up!" We got back in our HUMVEE and headed towards outpost number 7. We were lethal and proficient. Just before the front entrance of outpost number 7, our HUMVEE ran out of gas, so Hat radioed to Sgt. Middleton and AB to continue entering outpost number 7 and rendezvous with SSG Lewis. SPC Lonny Johnson, and PFC Royster posted guard as Hat and I refueled. When we arrived at outpost number 7, SSG Lewis was waiting on us. He said he'd seen the five positions where the insurgents were firing on outpost number 7 while he was driving in, so he split his squad up and told them to fire on the insurgents' locations with the AT4's and M203 grenades launchers, knocking out all five positions. I guess the insurgents thought the sandstorm gave them the perfect opportunity to strike the outpost, but they forgot about us quick reaction guys. Hat told SSG Lewis to check the bodies of the insurgents and return to base, and we headed back to the scene of our own firefight to check the dead for intelligence. Once we dismounted, AB and SPC Clayton Barnett took up security at the beginning of the street, and SPC Benjamin Saint James and PFC David Cousins secured the opposite end of the street. HUMVEE 3 searched the left side of the street and HUMVEE 2 the right. I stayed with Hat at the vehicle while he radioed back to the TOC. About a minute had passed when shots rang out on the left side of the street. Hat and I ran towards the sound of the shots, and found SPC Johnson, the crazy guy from New Jersey, shooting dead twitching bodies. As Hat walked up to him, Johnson fired his 9mm pistol and blood spray from the dead body splattered onto Hat.

Hat grabbed the pistol from Johnson and told him to check the bodies. Not knowing what to do, PFC Royster had taken up a defensive position, thinking someone was firing at him and Johnson, so Hat grabbed Royster too, and told him to move out. Royster looked back at Hat and said, "SSG Hadley, you have blood on your cheek!"

Hat kind of froze, and stared off into space while he wiped the blood from his cheek...

Hat's Story

The first time I had blood spray on my cheek was the first time I saw real blood. I was seven years old. Once a month, my mother and Aunt Lacy would have card parties where smoke filled the room, and the music was always blasting. Both my Mom and Aunt liked LL Cool J. We had a little routine going; Momma would say, "I'm gonna Knock you out, who's gonna knock you out?"

"I'm gonna knock you out!" would be my reply. Mary J Blige was also playing a lot; my aunt liked to think she could sing like Mary.

My job at the card parties involved cleaning the ash trays and handing people the drinks Aunt Lacy made for them. The card games they usually played were Tonk, Blackjack and Spades, but the Tonk table always had the most money - and the most drinks. One night at the Tonk table a man who had his son with him gave me a dollar to place his drink on

the table at his left side. As I circled around the chair to place his drink on his left side, the man giving the orders yelled out to the guy across from him, "Stop cheating!" The other man replied, "Fuck you!" the first man then reached under his chair and pulled a knife hidden in his

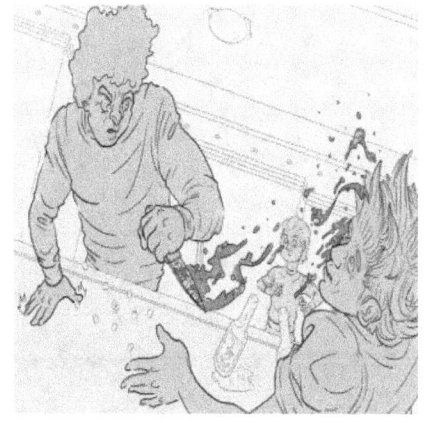

sock and with one quick motion cut the cheek of the other man. I was standing close to the man with the bleeding cheek, and the blood came pouring out between his fingers as he held his face with his hands.

I think the man that did the cutting must have been named Moses, because other men at the different tables said, "Oh Moses!" His name could have been Oh Shit as well though, because I heard a lot of that also. Moses grabbed the guy by the shirt and asked him if he wanted to get poked. I saw that the man with the slashed cheek   had a small pistol in his sock, and he was trying to reach for it when Aunt Lacy broke up the party with a long, black double-barreled shotgun she had behind the bar. My Mother grabbed me as Aunt Lacy yelled, "BOTH you motherfuckers are going to get shot!" Moses seemed

to take my Aunt seriously and he went over to hug my Aunt, but she just put that rifle right under his chin.

My Mother threw a towel at the man with the bleeding cheek, and then she took me to the bathroom and wiped the blood from my cheek and hair. Aunt Lacy came in and asked my mother if I was alright, but my Mother just replied, "Get that man out of my apartment!"

My Aunt asked me, "You okay, champ?" and tapped me on my head, and I nodded North and South and gave her a fake smile.

Things seemed to calm down. My Mother gave me a bath. Sitting on the edge of the tub my mother told me, "Bad people do bad things, and when you see trouble, you just move away from it."

I did not say much as my mom put me in bed with my little Sister and my younger cousin, Aunt Lacy's only child. I wanted to go to sleep, but as my mother closed the bedroom door, I heard a voice say, "Let's find that sumbitch and throw him off the roof!"

The voice belonged to Moses and our apartment was on the 16th floor - the top floor!

The thought of a body being thrown off the roof scared me, the blood on my cheek scared me. I cried myself to sleep with fright. I was woken hours later by a scream that came from the living room which sounded like my mother. When I entered the living room, she was hanging up the phone, crying and rushing around looking for her keys. She kept saying, "Stay in the house; Stay in the house!"

I said "Momma, I AM in the house," but she didn't listen. She finally found her keys and while putting on her sneakers, she told me to lock the door behind her and then she just ran out the door. I didn't know what to do, so I locked the door and sat on the couch waiting for her return. I looked out the window a couple of times, but it was pitch black. I drifted off to sleep on that couch, not knowing if my mother would ever return.

I woke up with the sun hitting my face and the sound of keys in the lock. I ran to the door as my mother walked in. She had blood on her shirt and pants, so I took a step back from her, looking her over just as she had done to me, hours ago. My Mother grabbed my hand and led me to the couch, sat down next to me, and told me Aunt Lacy was with God now. I thought about her statement for five seconds and said, "is she coming back?" just like that she told me Aunt Lacy was with God now. I thought about this a little. "Mom," I asked, "is she coming back?"

"No," she said, smiling at me, "God needed an angel, so he took your aunt."

I asked if she thought God needed a mom, and she smiled again and said "no." Then she looked me in the eyes, and I saw hers were full of tears as she said; "Now you need to take care of your sister and your cousin. You're the oldest, and I'm counting on you."

I told her I would take care of them and her too, and she told me to go back to bed. I thought she needed a kiss, but I didn't want to get any blood on my pajamas, so I grabbed her face with my hands and kissed her on the forehead.

Aunt Lacy's funeral was five days later. The funeral home was jammed packed. My Aunt knew a lot of people; my friend Rick and his family were there, and my friend Marty and his family were there as well. That man Moses, and his son, attended the funeral. Moses was crying hard just like my mother. The casket was closed, so the last time I saw my Aunt Lacy turned out to be in that bathroom when she'd tapped me on my head and called me Champ.

Hat was still wiping his face, when I heard shots and yelled, "Hat, Lonny is at it again!" with a grin.

"Let's mount up!" Hat responded as he returned from his pause. The quick reaction force had done their job and was heading back to Camp Slayer. Just as we entered the camp, we ran out of gas again, and again Lonny Johnson and Royster took up fighting positions as Hat and I refueled. Back in the tent we drank a lot of water, refitted, cleaned our weapons, took care of personal hygiene, and slept.

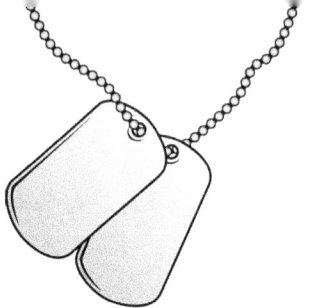

CHAPTER TWO

After six straight hours of sleep, which is a record for us, I woke up to find the supply guys had dropped off 8 cases of Gatorade in our tent while we were sleeping – none of us heard a thing. The guys started stirring as those Gatorade bottles had us drinking our own saliva in anticipation of gulping it down, but just at that moment Hat walked in.

"We're running rear security on a day mission, so gather round," he said. We held hands and Hat told AB it was his turn.

We bowed our heads and AB said, "Dear God, thank you for this day. Send your angels to protect my wife and son Manny and bless everyone here's family. Thank you for helping me turn from my wicked ways and bad thoughts about white people and thank you for Jennifer Lopez. Amen."

"Jennifer Lopez! and white people? What am I? a blue avatar person" said Benjamin (who is white and AB's best friend) while slapping the back of AB's Kevlar helmet. SGT Middleton laughing said "your prayer turn is revoked" as we all walked laughing.

Hat had made us all pray every morning after Utah got his legs blown off. The only condition for praying was that we could not pray for each other's' safety, which did not make sense to me. I believe in GOD, but I am not a praying person, but when it's my turn I pray, I pray. The last time I prayed other than my turns in Iraq was when I was 14 years old. My friends and I killed a deer during off season. My Father, receiving that citation letter from Minnesota Game and Fishing told me he was going to punish me. I remember it now; I said, "God, please do not let my Daddy punish me!" Unfortunately, God did not listen, and my Dad tore my backside up!

"What about the Gatorade?" I asked Hat. He told me and PFC Royster to put the Gatorade bottles in ice, and that was that.

Sgt. Middleton yelled "Perform your combat checks and hit the road in 10 minutes!"

Rear security is boring but safe. We often pulled rear security after combat killing to reduce the stress level. We gassed up and rolled to the ready station line just inside the gate. We ran communication checks, conducted preventive maintenance checks and services on the vehicles, and cleaned our weapons again. Hat and SGT. Middleton went to the

TOC and listened intensely as the fire fight played out on the radio. Rear security sucks: it's a "hurry up and wait" game, but vital due to the insurgents' constant efforts to attack the rear. After waiting for five hours, we were informed to stand down and return to our squad area.

We ate lunch and then we went to the weightlifting tent, certain of one thing: music wars. There is always a war in the weightlifting tent. This war was between Rap - represented by SPC Lonny Johnson, and me - the Rock'n'Roll king. We had two music boxes at the opposite ends of the weightlifting tent. Lonny would take his music up a level if he thought my music was too high. Today he did the usual, and, so in return, I turned up the volume on my Green Day song. I turned up the volume on my Green Day Song. Lonny then turned up his music on a Jay-Z song. After the back-and-forth encounter, Lonny got in my face and tried to use his rank, "Back down PFC, turn your music down, I out rank you! Lonny said. But I told him, "The hell with rank! Fuck you!" Now while this going on, the other guys would be dancing to the music while waiting their turn on the weights.

Sgt. Middleton, who was lifting weights with Hat, broke up the argument but words continued to fly between me and Lonny. No one

noticed, Hat walking to the music box. Hat put in his Toby Keith CD, told SGT Middleton to cut the music from the other music box and said, "You motherfuckers better sing…"

I've counted up the cost, I know the sacrifice,
Oh, and I don't want to die for you,
But if dying's asked of me,
I'll bear that cross with honor,
'Cause freedom don't come free.
I'm an American soldier, an American,
Beside my brothers and my Sisters I'd proudly take a stand, When liberty's in jeopardy I will always do what's right, I'm out here on the front lines, so sleep in peace tonight. (Toby Keith)

Lonny whispered to me, "You see what you did, I hate this damn song and that damn Country and Western shit!"

Hat looked pointedly at him, and Lonny stopped talking and sang louder because he knew why Hat had this ritual.

Hat and Utah used to fight over Rap and Country. Hat had an old school head for rap music, and liked Jay-Z, Run DMC, anything by LL Cool J, Nas, Father MC, Fabolous, M.O.P., Mos Def, MC Lyte, Mobb Deep, The Lox, Foxy Brown, Biggie, Nice and Smooth, Public Enemy, BDP; basically, anything before the year 2000. Utah was Country music all the way, and when Utah got his legs blown off during his second tour – which was my first in Iraq - Hat had to pack Utah's things up and ship them back to Fort Bragg. For some reason best known to him, Hat never did pack and send Utah's "Best of Toby Keith" CD.

After dinner and a cold shower everyone started to drink the cold Gatorade, now in cool water. Royster started talking about basketball and Michael Jordan. Royster started telling us how great Jordan was, like none of us had seen him play. Joey Pisciotta was arguing, "Hey, hey, hey, the black mamba is retired, but he was noooo joke, Kobe's the man." we know but he's retired; Kobe's the man."

PFC Royster said, "Kobe has five rings, Jordan my idol has six." AB said, "you are a fool! YOUR IDOL YOUR IDOL," in his heavy Latin, New Mexico New York City accent, "I did not see Jordan when you were getting shot at; I did not see Jordan cover your 6 during the firefight either. Your idol, green boot is that man over their sleep!" while pointing at Hat. UR IDOL YOUR IDOL," in his heavy Latin, New Mexico accent, "I didn't see Jordan when you were getting shot at; I didn't see Jordan cover your 6 during the firefight either. Your idol, green boot is that man over their sleep!" while pointing at Hat. "Michael Jordan my ass! You better think Kabron, Michael ain't doing shit around here. What? Are your boots made by Nike? Everyone laughed loudly waking Hat. Angrily, Hat told everyone to shut up, read your mail and go to sleep. Then he told Royster to step outside the tent and go into the bunker.

Hat kicking two rats away as he stepped into the bunker said to Royster "Who raised you, feed you, clothed you, gave you money for college, wiped your nasty nose, stayed with you when you were sick,

taught you how to tie your shoes, wiped your ass, taught you how to read and write?" "My mother" Royster replied. "Now when you're playing basketball who do you want to play like? emulate?" Hat asked. Royster replied "Jordan." Hat said "Don't believe that media bullshit, ask Jordan who he wanted to play like, and he'll probably say Doctor J, ask Jordan who his role model is, and he will say his father why, well, because he saw his father go to work daily and come home. How often did you watch Jordan daily? Do commercials and the basketball season make him a role model for you? Or does your mothers' work ethic and daily interaction inspire you? When you graduated from basic training and airborne school who attended and cried with pride? The media will take the next great basketball player and label him a role model and as soon as the role model makes a mistake, that same media will attempt to destroy them. Do you smell what I am breathing? Don't let the media determine your role model, get the picture? Besides, Lebron and Curry are my dudes. They will probably topple every basketball record." As they entered back into the tent, AB and PFC David Cousins were arguing. PFC Cousins received a picture from his grandfather in front of a flag flown in their yard. PFC Cousins told AB his grandfather said, the flag flown in their yard was the new American flag. AB looking at the picture and yelled "THERE IS ONLY ONE AMERICAN FLAG AND THIS SHIT AIN'T IT." PFC Cousins seeing Hat walk into the tent walked quickly to Hat and said, "my grandfather sent me this picture and said this is the new American flag." Hat looked at the picture and told PFC Cousins "Your grandfather is a fuckin moron." Hat walks to rucksack, pulls out his American flag and states "there is only one American

flag, the American flag does not have a thin blue line, thin red line, thin yellow line, or thin black line. That shit pointing to the picture is a decoration. When you were a kid and said The Pledge of Allegiance, this was the flag (shaking the American flag in his hands). When you raised your right hand to take the oath of enlistment, this was the flag (shaking the American flag). You swore that you will support and defend the Constitution of the United States against all enemies foreign and domestic (shaking the American flag) this was posted close to you (again shaking the American flag). Again, that flag in your grandfather's yard is a decoration. AB, pull out that Puerto Rican flag you carry with you." AB goes to his rucksack and pulls out the American flag and then the Puerto Rican flag. Hat with the Puerto Rican flag in hand ask AB, "where are you from," AB replied" South Bronx." Hat said "what are you? AB replied, "I am a No Limit motherfuckin American Bad Ass Soldier from Puerto Rico by way of the South Bronx and New Mexico." Hat said, "What would you said if you saw the Puerto Rican flag denigrated with a thin blue, red, yellow, or black line," AB replied, "I'd piss on the flag and slap the Kabron displaying it." Hat looking into the eyes of PFC Cousins said "the American flag is over 240 years old. The colors are red, white, and blue with 50 stars and 13 alternating stripes, seven red, six white. These colors NEVER run. Nothing added, nothing taken away. No thin blue, yellow, black, or red line." Hat's voice softens as he puts his hand on the shoulder of PFC Cousins and said "When you get home, and you will get home, give your grandfather my American flag (Hat hands PFC Cousins his flag). Should your grandfather ever become confused again, tell him to google 'Raising the flag on Iwo Jima'. Now GO….TO…. SLEEP" SGT Middleton and SGT Pisciotta, slap PFC Cousins on the back of his head and then SGT Middleton yells "Donald Gordon" and everyone including PFC Cousins replies "Donald Gordon." SGT Middleton and SGT Pisciotta go over to PFC Cousins bunk and show him how to correctly fold the American Flag. AB joins them shortly after he folds both his flags. Once completed; PFC Cousins placed the American flag in his rucksack. About two hours later I woke up because my bladder was full and I need to urinate, so as I prepared to take that long walk to the latrine, I turned to see AB urinating in an empty Gatorade bottle. I walked over to him and said, "Why are you pissing in the bottle?"

AB said in a quiet voice, "The latrine is too damn far, I gotta go now." Well, I thought, I gotta urine also - so I urinated into my own empty Gatorade bottle, placed the bottle under my bunk and went back to sleep.

About an hour later, the tent was shaken by a loud yell. "YOU MOTHERFUCKER; YOU MOTHERFUCKER!" SPC. Benjamin Saint James was yelling.

"What?" AB yelled, as Benjamin threw the bottle of Gatorade urine at AB.

AB jumped up and punched the hell out of Benjamin and they went at it. Fists were flying until Hat woke up and got in between them, yelling, "What the fuck over?"

Benjamin said, "I wake up thirsty and reach on the floor for my bottle of Gatorade, and when I drank it, it was piss!"

I started laughing inside and Hat said, "What? Well, how do you know AB did it?"

Benjamin said, "Because his bunk's next to mine and AB is a lazy motherfucker."

Hat turned to AB and said, "What happened?"

"Well, I needed to piss so I used my empty Gatorade bottle instead of hitting the latrine; that damn latrine is way too far, and I had to go badly. Benjamin reached over with his long-ass arms and grabbed my Gatorade bottle full of piss and drank it!"

Hat started laughing, and looked at Sgt. Middleton, who pushed Hat to the side as fists flew again. Benjamin threw AB to the floor and grabbed the bottle of piss and poured it on AB. In no time, those damn fools were rolling on the floor, but nobody in the squad was breaking up the fight because they smelled like urine. Hat finally broke up the fight and told Benjamin and AB to get a mop and pine oil and clean up the mess.

At that moment, a private came in and told Hat to report to the TOC. We all cleaned up the tent together and took showers. Then SGT Pisciotta told us to clean our weapons – AGAIN – Hat came back from his briefing at the TOC. He called Sgt. Middleton outside and told him we were going on a daytime operation. Hat said, "Intelligence informed us of potential Al-Qaida in a village about 20 kilometers away. We'll have Apache support on the ready, and two squads will surround the village; another squad will back us up. We'll be doing a dismounted village assault; I need you to brief AB, Benjamin, and Pisciotta. We'll switch up for this dismount; I'll take SPC Barnett, PFC Royster and PFC Frost." We barely received a couple of hours of sleep last night but that was normal.

On dismounted assaults we took extra extra rounds and left the water in the HUMVEEs. When I asked Hat why we left the water, Hat told me he saw a movie called 'The Sands of Iwo Jima' where some Marines were in a firefight and ran out of bullets, so they sent this guy back to get more ammo. The guy sat and drank coffee while his fellow Marines died. Hat said that personally he'd rather send somebody for water than ammo, so we always loaded up on ammo and removed our canteens.

Dismounted door-to-door break-ins are serious business because they are slow, painful, and done in close quarters. In Hat's first two tours in Iraq, door to door included explosives. Blowing the doors off was the norm. Hat now uses a special door gadget to knock the doors

in; he did not like what the explosives did to the Iraqi homes. Having two green soldiers with us that day made us more careful, especially as bullets were coming at us intermittently from homes that were close to each other. As we closed in on the home that intelligence believed housed Al-Qaida, my heart rate increased.

Sometimes intelligence is off by a home or two, but today they were spot on. What they didn't tell us, however, was that the adjacent homes housed Al-Qaida also.

The interpreter kept telling them to surrender or be killed, but nobody was surrendering. Door-to-door is always a pain because you must place yourself in a position where you shoot and place yourself in a position to get shot.

Once inside a home, you're trained to shoot anything moving; your trigger finger is always at the ready.

After clearing a home, Hat exited out from the front door and then kicked down the door of the adjacent home. After he kicked down the door, a sniper on a rooftop shot SSG Hadley.

He fell forward into the doorway, and as Barnett and Royster came to pull him fully inside the home, at the same moment, an Al-Qaida man came bursting out of the sheet rock.

The two green soldiers had placed their weapons on the floor to have both hands free to assist Hat and even Hat's weapon was not at the ready, so when the Al-Qaida burst through the false wall, the green soldiers had to scramble to get their weapons. The Al-Qaida guy obviously had a beat on them, and he was just positioning himself to fire

when Hat pulled out his 'if I'm ever captured 32 revolver pistol and shot him in the head. The Al-Qaida fell dead.

The green soldiers had their weapons sorted now, and I heard Hat tell them to shoot up the place. I was on the other side of the room calling for a medic, but when I saw what happened I started shooting right at those walls as well. We killed four more Al-Qaida hiding in those walls. AB shot the sniper who'd shot Hat, and as it turned out, Hat had his vest on and was OK, so he told us to check the bodies. He then got on the wireless communication and told Sgt. Middleton to re-enter the homes that had already been cleared and shoot all the walls. More Al-Qaida was found in those walls. I thought these snipers al-

ways aim for the head, how is Hat still alive?" Barnett said, "I heard while we were "locked in before deployment" and told not to exceed more than 25 miles from base, Hat and SGT Middleton drove 25 hours straight to South

Dakota to meet with a Medicine Woman at the foot of Mount Rushmore. Half of Hat's Kin Folk are Blackfoot Indians and before every deployment Hat and SGT Middleton would meet with this little old Indian Medicine woman. She would pray for them."

SGT Middleton spoke up and said, "Years ago, Hat's family flew to South Dakota for a funeral. Hat's great grandfather was 92 years old. He told Hat that peace comes from within. South Dakota's Blackfoot Mountains is a place where you will always find peace, especially in those mountains. Prior to our first tour in Iraq, we drove to South Dakota. We stopped at a gift store where we saw this little old Indian woman selling gifts outside a gift store. She made eye contact with us and pointed us towards the back of the gift store. She held our hand and prayed for us. We continue that tradition prior to every deployment no matter if we break the mileage restrictions or not." Barnett and Royster still upset over the battle and seeing Hat get shoot said, "I am going on the next trip to South Dakota."

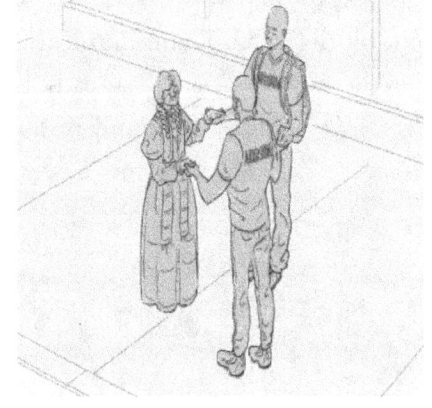

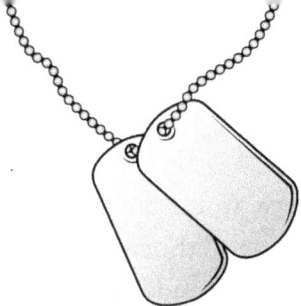

CHAPTER 3

Sgt. Middleton ordered crazy Lonny to search one home and Benjamin and I to search the bodies for any kind of intelligence in the adjacent home. I was searching bodies in the second house when shots rang out from another home.

Benjamin and I both came running into the home where the shots were heard, and Lonny said, "Some of the bodies were still moving." We continued our search and found maps, papers, laptops, and cell phones, all of which would be turned over to intelligence.

I noticed Hat looking at the dead Al-Qaida, he had a certain look about him. Was he thinking about the man he shot or Utah, I wondered? Sgt. Middleton had once told me that after the first tour, Utah smuggled three 32 revolver pistols back to Iraq in his A-bag. Utah gave Hat and Sgt. Middleton a revolver each, they made a pack to shoot each other in the head prior to ever being captured by Al-Qaida. Maybe Hat was thinking that he almost lost his life, I'm not sure, but he had a strange look on his face. Finally, he said, "I wish Rick was here. When we used to get intelligence for a mission, Rick would dismount about 150 yards from the town, put on an abaya, and sneak off to take pictures of the location we would attack the night before we moved in. Rick would take pictures of all the entrance points and surrounding areas, but sometimes the shots would be messed up, because Rick said it was hard getting a direct picture." Then suddenly, Hat said, "I've got a headache," and he ordered everyone back to the HUMVEE's where he told us to drink water. It would be nightfall before we finally mopped up and were heading back to the camp. Now in the HUMVEE's the guys started to talk. Hat was very quiet for a while. I don't think he ever thought he would get shot, so no one said a word to him, to include Sgt. Middleton.

Hat told Sgt. Middleton to ride back with him, and on the way back I heard Hat say, "Marty, I went too far. I should have waited until the other squad cleared the high ground; we all know Al-Qaida likes to take advantage of the terrain."

Sgt. Middleton said, "Don't worry about it," but Hat didn't answer. A little later I heard Sgt. Middleton ask him "How's your headache?" Hat did not answer him.

When we arrived back at base, we did an after-action report with Captain Ron Ward, our Company Commander, then stood down.

PFC Royster, not wanting to bother Hat, asked Sgt. Middleton, "Who was Utah?"

Sgt. Middleton thought for a minute then replied, "Utah just saved Hat's life, Royster, and he isn't even here. Utah gave Hat and I revolvers during our second Tour, because he was scared after we all heard about the beheadings during our first Tour. Utah didn't want anyone to be alive if we ever got captured."

"Where is Utah now?" Royster asked.

"Shut up! Stop talking about Utah, Royster!" As Pisciotta hissed, but Sgt. Middleton didn't seem to mind continuing.

"During our second Tour, Utah got his legs blown off by stepping on a land mine by the side of the road, and NO, he is not dead!" Sgt. Middleton continued, "I talk with him often; he has prosthetic legs."

The next day started like the rest. We would gather for morning prayer, and Hat said today it was his turn to pray. Stretching out his hands for us to join hands together, he said, "Lord God, we come before you with bowed heads and humbled hearts, thanking you for another day. We ask for forgiveness for our sins, God, and ask that you take care of all our loved ones. Father, we ask that you keep our families safe; surround them with angels to protect them. Give our families strength; gird them up with the Holy Spirit. Let us all say Amen."

At this point we would usually fist five and move out, but today Sgt. Middleton interrupted Hat's prayer at the end and said, "God keep us safe!"

Joey Pisciotta broke hand contact and said, "Man, what are you saying?"

"Man, we are going to have a bad day!" Lonny added.

"We never-ever ever EVER pray for our safety!" Lonny said.

Benjamin said, "Man, come on, let's pray again."

Lonny kept repeating, "Man, oh man, we are going to have a bad day."

Benjamin said again "Let's pray again" AB started to pray in Spanish while Royster and Barnett grabbed AB's hand.

Finally, Hat said, "ENOUGH!" and pulled Sgt. Middleton outside, but I could still hear their conversation with all the commotion inside the tent. "We have four days before we rotate back to Ft. Bragg, why lift the lid on a boiling pot?"

Sgt Middleton replied, "Man, you got shot yesterday; we cannot STOP REWIND AND EDIT! I couldn't sleep thinking about that!"

"Look, we just need to think about the men…" Hat told him, but Sgt. Middleton cut him short.

"Donald Gordon," Sgt. Middleton replied quietly, "Donald Gordon."

Convinced the squad was in disarray, Hat walked back inside the tent and told everyone to stop arguing and sit down. He then proceeded to tell us all a story about his second tour in Iraq.

"One night on my second tour in Iraq, a runner from the TOC came to our tent and told me First Lieutenant Ward wanted me at the TOC. When I arrived at the TOC, 1LT Ward waved at me to come close, and as I got close, I could hear a platoon of Soldiers being ambushed by insurgents on the radio. The insurgents had one of our radios and broke into our network. I listened as the ambush platoon was hit with vehicle borne IED'S and roadside IED's all at the same time. Just before the platoon was hit, the insurgents came on the communication network and said, "You American pigs, you infidels; listen as your comrades die!" "Less than 10 seconds later I heard calls for help from Soldiers shouting, "WE HAVE MEN DOWN!" I listened in horror as the insurgents came back on the network and said, "Yes, come and help them!" then we heard the sounds of shots coming from AK 47's…then the insurgents said, "You see, you see what we can do to you?" all in broken English and with an Arabic overtone. "They said, "Go home now or go home bloody, but you will go home." I heard the buzz of Blackhawk helicopters. The pilot said he could see 15 American Soldiers dead." "I went back to my bunk and thought, and then for some reason I thought of Dirty Burt." Everyone looked at Royster, and on queue Royster said, "Who is Dirty Burt?" while everyone laughed with anticipation. "Dirty Burt is Bartholomew Jackson," Hat replied. "This guy was always around in our neighborhood when I was a kid. He often wore a dirty old field jacket from the Vietnam era, an Air Force baseball cap, black military jump boots, dirty jeans and a green T-shirt, no matter the weather even in the summertime. His face was black and dirty, and his beard was a mess. His eyes were glassy, but what I noticed more than anything was his teeth were white and his fingernails were groomed. He always had a nail file in his hand. For some reason when I was 13, I sat at the end of the park bench where Dirty Burt was, and as usual he was constantly talking to himself and making weird noises. I was waiting for Marty to come out and play. I strained my ear to listen to what Dirty Burt was saying to himself, and after a while I slowly inched close to him, just to listen. As I moved close," Dirty Burt said, "You Alice's boy, right?" Yes, "I said," and he said, "You want to learn something," but just then Marty – Sgt. Middleton - walked up. As I turned to look at Marty, he had this frightened look on his face. When I turned back to look at Dirty Burt, he moved up and sat next to me. Marty had the basketball in his hand and said, "Let's go," but Dirty Burt got up and moved to a dirt spot in the grass and started drawing something. Some older kids came by and Leon their leader teased, "Clean yourself up, Dirty Burt," throwing his can of coke at him. Leon's friends laughed. Marty and I stood ready to

fight if they came our way. Leon and his stupid friends kept going and never stopped to mess with us, and for his part, Dirty Burt kept right on writing in the dirt.

As I moved closer leaning over Dirty Burt's shoulder, I saw this: Hat got out a piece of paper and started to write.

	1	2	3	4	5
1	A	B	C	D	E
2	F	G	H	I	JK
3	L	M	N	O	P
4	Q	R	S	T	U
5	V	W	X	Y	Z

Dirty Burt said, "Tap Tap Tap – Tap = L... and Tap Tap – Tap Tap Tap = H," Hat continued, "And instead of playing basketball that afternoon, we learned the Tap Code from an Air Force pilot who flew food supply missions into Vietnam in 1972 as a 21-year-old. Dirty Burt told us he learned the Tap Code as a Prisoner of war in Vietnam."

Sgt. Middleton chimed in, "Before the day ended, Hat and I would substitute the tap code with numbers, so instead of tapping, I would say, "32," and Hat would reply, "M," I would say "11," Hat would reply with, "A," and so on. 42 was R, 44 T, 54 Y and so on. Hell, we played spelling games all day!"

PFC Royster said, "That's the numbering Tap Code we use to communicate in our sidekicks!" and Hat and Sgt. Middleton smiled at each other.

Hat paused for three to four seconds, as if he blanked out, then he continued, "Marty and I would see someone we both knew and would number their name. I even taught the numbering code to my sister and Cousin One.

"So how did we get the sidekicks?" Asked Royster

Hat replied, "Well, after I got back that night from the TOC, I woke up SPC Jerry Bernard (JB) and SPC Hiram Maurisee. Those two boneheads joined the Army just before 9/11 to pay for their college bills, and then they got caught up in the war. Both were engineers and computer geeks. I told them what happened in the TOC and explained the numbering Tap Code. Both were big thinkers; not much on killing. A week later they had changed the sidekicks you have into the numbering code."

"What about the Terrain Recall Glasses we wear?" Royster asked. "They're simple sunglasses that had an MP3 player on them," Hat replied. "These two guys put video recording devices and imagery in the place of the MP3 player, so now we can record and detect terrain changes from day to day."

"So, what happened to them?" Royster questioned.

"Well, after they perfected the sidekicks and the Terrain Recall Glasses, 1LT Ward summoned me, JB, and Maurisee to the TOC, where we were greeted by a grinning full bird Colonel by the name of Colonel Robert Irvin. I knew he wasn't in our chain of command; we shook his hand as the 1LT made the introductions." The 1LT said, "Sir, these are the two hard charging soldiers who invented the

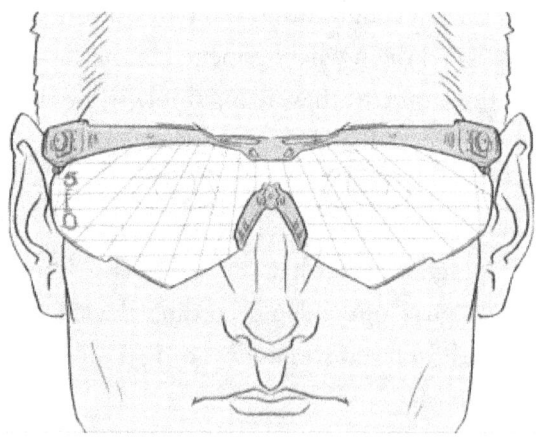

numbering sidekick, and the Terrain Recall Glasses," pulling on the arms of Maurisee and JB as he introduced them.

"Both of them seemed embarrassed, but I was proud - and I was even prouder when I heard Colonel Irvin say, "I want you two to return to Ft. Bragg and report to the emerging Technology cell of Psychological Warfare Operations (PSYOPS). Both JB and Maurisee attempted to explain to Colonel Irvin it was my idea, but I interrupted their interruptions and said, "Sir, these men will be on the first thing smoking!" Then I pulled them out of the TOC."

"JB and Maurisee grabbed me and thanked me. For most of our maneuvers I kept them as rear guards. Both men were trigger pullers, but they lacked the killer instinct."

Suddenly, a runner for Captain Ward summoned Hat to the TOC, so Hat did not get to finish his story. When Hat returned, he said, "Four days left, and we're delivery boys. No bad day for us, Specialist Johnson!"

Johnson muttered sourly, "The day is just beginning," but Hat explained the details of our mission and, after conducting pre-combat checks, we were off to deliver Thuraya Satellite Phones.

As we approached a road intersection on the route, someone down the street fired off an AK-47. We responded unaware that another military patrol was in the area. Prior to entering the town, Hat had Benjamin and AB dismount and climb onto a roof, and Sgt. Middleton called the TOC and informed them about the shots fired. The TOC called the Joint Intelligence Operation Center (JIOC), who had sophisticated blue force tracking devices implanted on all military vehicles. It was a good thing to have those devices, because the JIOC quickly assessed we were about to engage a Platoon of Marines! The JIOC contacted both TOCs in a matter of seconds and just managed to call off a confrontation between friendly forces. Hat told us this ability had been missing a year ago, as he often heard of fratricide back then.

Sgt. Middleton used his sidekick to call AB and Benjamin, who had made their way on the roofs of houses. AB sent a message back to Sgt. Middleton, who joined them both on the roof where they found AK-47 shells, along with a lot of cigarette butts. From that position on the roof, you were able to see both military units approaching the town - one from the East and one unit from the West. If the JIOC hadn't been functional, the insurgents would have created an environment in which we would have engaged each other.

Sgt. Middleton, trying to redeem himself for praying for our safety, said to SPC Johnson, "Not bad, Johnson," but SPC Johnson just replied, "The day is not over."

I looked at our GPS and noticed we were only five miles from outpost number 6, where we had to deliver the Thuraya cell phones, and we continued our mission. Four miles from the outpost number 6, two motorcycle riders carrying another passenger with AK-47's started circling our three vehicles. Sgt. Middleton and Benjamin used their sidekicks to inform Hat, and when Hat saw those motorcycle guys, he told Joey Pisciotta to shoot the bike on our left and for Lonny to shoot the motorcycle bike on our right. Unfortunately, telling Lonny to shoot the bike is like telling Lonny to shoot the guys, so all four guys crashed and burned!

Hat, wearing the Terrain Recall Glasses, was attempting to recall this route when he decided to stop the convoy. We dismounted and went into a defensive posture. Hat kind of froze for a second, and then for some reason he requested air support; the TOC must have thought it was an unusual request because we were not in a firefight, but they sent two Apache helicopters anyway. Hat must have seen something before, because he directed the pilots on a route ahead of us on the

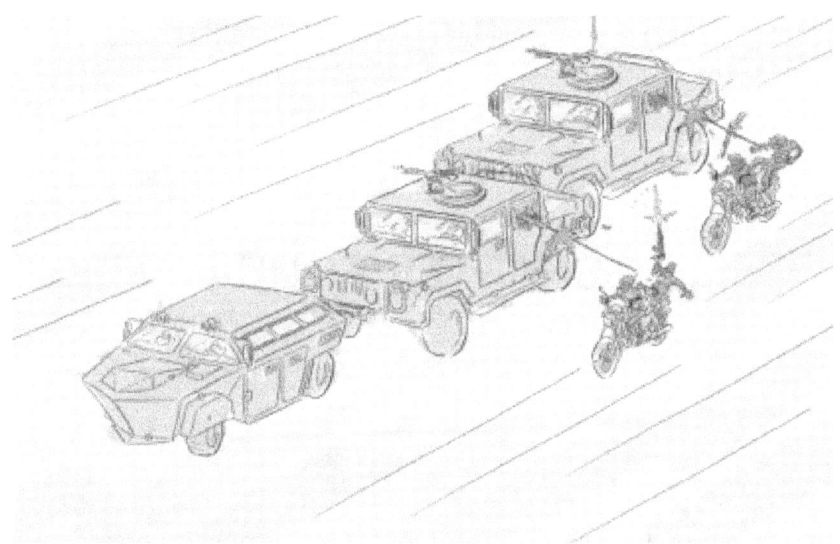

road. About a mile down the road, the Apache helicopters blew up three VBIED'S and four IED's, and we all cheered out loud as we saw the black smoke.

Hat said the motorcycle guys were riding around us trying to divert our attention onto them while we headed into a trap. It was a good thing those terrain recall glasses work. After a quick sigh of relief, we proceeded to outpost number 6, delivered the Thuraya Satellite phones, and returned to camp all in one piece. Even SPC Lonny Johnson smiled.

It was considered bad luck to pray for ourselves, but so far, we had escaped danger. With only four days left in Iraq, and six hours remaining in this day and with no mission in sight, we all sat on our bunks, tired, but relieved to know we had probably just conducted our last mission on this tour.

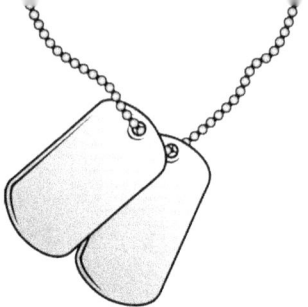

CHAPTER FOUR

I looked around at the guys and thought about how we prayed every morning, how we'd developed as a team. Man, we were more than a band of brothers, we were family. My eyes were looking for Hat, but he was not in the tent, so my eyes turned to Sgt. Middleton. SSG Hat and SGT Middleton took pride in keeping everyone alive; these combat veterans were professional Soldiers. I figured it was odd that we'd been back for 10 minutes, and I hadn't heard SGT Middleton say, "Clean

your weapon," yet. I stood up to get the cleaning tools for my weapon when suddenly Hat dragged in one of the cooks from the Dining Facility (DFAC). He had him in a head lock.

Hat screamed to Royster, "Go get the captain!" Telling Lonny to grab a chair, Hat kept tight hold of the cook. Everyone was stunned. Lonny got a chair and Hat placed the cook in the chair, telling Lonny to hold him down, then Hat looked at me and told me to duct tape him to the chair.

As I was taping the little Iraqi man, I looked into his eyes. His legs and arms were shaking, and he seemed very scared, maybe because Lonny was giving him love taps across his face.

The Iraqi man started talking in Arabic, and Lonny told him, "Shut the fuck up! "And snatched the duct tape from my hand and taped his mouth.

Sgt. Middleton raced from across the tent and asked Hat, "What the fuck happened? What the fuck … OK, calm down everyone."

Hat said "I was on my way to the latrine when I saw this guy pacing, counting, and writing stuff on a piece of paper with this pen. When he saw me, he stopped and became nervous? So, I walked up to him, grabbed him, put him in a choke hold and dragged him in here."

"This motherfucker works in the DFAC," Lonny said, giving him another slap. The little Iraqi man was crying. I checked his pockets and found he did indeed have notes written in Arabic. Lonny was ramped up and grabbed his weapon, but Benjamin and AB took his weapon from him. Hat told Joey to find Royster and the Captain. As Joey exited the tent he turned straight back around as Captain Ward, PFC Royster, and an interpreter entered the tent.

Hat began to tell the interpreter and Captain Ward what happened. The interpreter looking at a slumped Iraqi man in the chair, walked over to check on the Iraqi man. The Iraqi man was slumped over as if he was dead. The interpreter looked at us and asked, "Did you guys kill him?" Lonny went over and slapped the Iraqi man's face real hard, and the Iraqi man woke up looking dazed and confused. Lonny said, "No Sir, we didn't kill him, this soft ass just passed out."

Captain Ward snapped at Lonny, "Cut the shit and cut the duct tape from around his mouth!" The interpreter was shouting at the Iraqi man in Arabic, but the cook didn't say anything, so the interpreter asked for a bucket of ice. Joey Pisciotta went to the DFAC to retrieve a bucket of ice, as the interpreter continued to yell at the Iraqi cook. The interpreter instructed Hat to remove the duct tape from around the Iraqi man's legs, which had been taped to the chair. Hat removed the tape, and the interpreter told Hat to remove the Iraqi's man pants. Hat looked at the interpreter in surprise. The interpreter winked at Hat, as Pisciotta returned with the bucket of ice.

"Stand him up and lower his pants to his knees, then place his right leg in the bucket of ice," the interpreter said. As we did it, the interpreter yelled at the Iraqi man, while the rest of us wondered what was going on. The interpreter took the papers found on the Iraqi man from Cpt. Ward's hand, and read out loud whatever was on the papers, and the Iraqi man started to cry.

so, the interpreter yelled some more at him in Arabic. The Iraqi man was sobbing, and then he looked down at his penis and began to urinate. No one was sure what was going on. The cook now very upset is crying profusely, crying, and speaking in Arabic. I started to feel sorry for the little guy as he spoke and cried. Finally, the interpreter told Captain Ward to release him and bring him to the TOC. Captain Ward still visibly mad with Lonny barked loudly "remove his leg from the bucket of ice and bring him to the TOC."

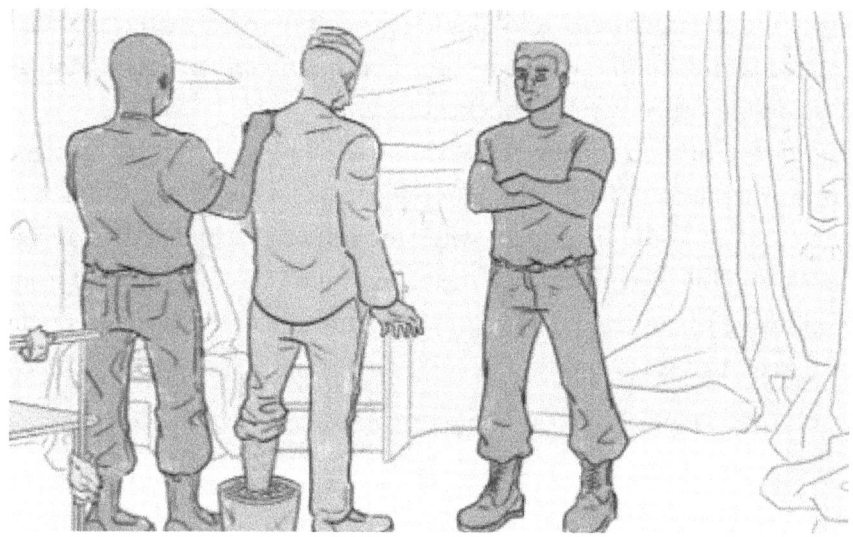

After the Captain left, Joey and Hat escorted the Iraqi man to the TOC. Royster asked, "What's with the ice?"

Lonny ignored him and grumbled, "I'm not sure why Hat didn't just choke him out and snap his neck, we wouldn't be wondering about this shit!"

AB shot him a dirty look and said, "So you didn't feel for this guy? Man, your heart is cold and rotten!"

"You damn right!" Lonny countered.

"But what's with the bucket of ice? It just doesn't make any sense," Royster repeated.

Even quiet SPC Barnett said, "Yeah, what's with the ice."

When Joey returned from the TOC, Joey explained, "The interpreter told the Iraqi man that if he didn't tell us the truth his leg would freeze, and his dick would fall off! That Iraqi was so scared he urinated on himself with embarrassment!"

Lonny and Royster laughed as Hat walked into the tent with Hat saying "Keep your laughter to yourself, dickheads, the cook is in trouble. His family is being held hostage by some insurgents who'd found out he was making a living by cooking for us. They raped his wife right in front of him, his son and daughter, and told him if he didn't return with the measurements from the DFAC to the Command Tent in steps, they were going to rape his daughter and son and kill them both."

Suddenly, our shock was replaced by anger as Hat told us the cook's story, and all at once the squad started cleaning their weapons and drinking Gatorade. I went back to the TOC with Hat, where Captain Ward told Hat he'd received the go ahead to conduct a mission that did not involve saving Americans. This mission was about saving an Iraqi man family.

In the tent, the only noise you heard was movement. Hot-head Lonny approached Hat, and in a soft voice he asked him, "How can insurgents arrive in a community, take a home, rape a woman and hold a family hostage without people in the community knowing? This is bullshit, but it's like this Hat, if you don't want me to kill these bastards, you better leave me here 'because somebody's going to take a long blink tonight you smell what I'm breathen?"

Hat looked Lonny in his eyes and simply tapped Lonny's weapon and said, "It's time for good people to do bad things."

Cpt. Ward, the interpreter, and the Iraqi cook walked into our tent. Lonny who had laughed and hit the cook stopped in front of the cook and shook his hand. Cpt. Ward briefed us, and he professed that the plan was a simple one. SSG Lewis' squad was backing us up. The insurgents had taken two homes next to each other, with four insurgents in each home. We were to secure the area around both homes, kick down the doors, rescue the family and get hostages for intelligence to interview. "In other words," Cpt. Ward said, "we need these folks alive, Soldiers!"

Lonny, who is a smartass, never said a word, he just smiled. Reinforcing what Captain Ward said, Hat told us to gather around and gave us our riding arrangements and asked whether we'd conducted pre-combat checks.

"HOOAH!" we responded.

"DONALD GORDON!" he announced at us, and "DONALD GORDON!" we replied loudly.

Captain Ward went over our route, and as it turned out, we even had imagery from an UAV. We received more intelligence for this one mission in less than an hour than we'd received for other missions that had been planned for weeks. On the ride out, SPC Barnett who is green and was told by Lonny, "Shut the fuck up, don't ask questions, and follow the leader!" when he first arrived at our squad, rode with AB and Benjamin. Barnett asked, "Why do we say Donald Gordon? And who is he?" AB told him, "We say 'Donald Gordon' for a couple of reasons; Hat is an old rap head, but he's also a student of military history. He believes in mottos. He can tell you about every war or conflict America has ever been in. He thinks America changed, we use to have a philosophy in war of maximum amount of fire power upon the objective, but now, we as Americans have gone soft; we do measured aggression now, we measure our response, the intent in our response is to send a clear message, but war is war Hat will tell you." Hat will tell you, when a motherfucker slaps you, you slap his ass five times harder and break his nose, That's American. Benjamin interrupted and started laughing, "This is why we say Donald Gordon. When Hat and Marty were 11 years old, they were playing basketball and Hat argued with his second-best friend Donald Gordon over the score of that game. Hat called Donald Gordon a liar, and Donald, who was bigger than Hat, said, "Let's just play!" but Hat insisted on calling him a fuckin liar. Hat then pushed Donald Gordon, and they had a fight and Donald Gordon beat the brakes off Hat. Hat was bloody and his eye was swollen." "Why didn't Sgt. Middleton jump in?" Barnett asked. "Because it was a fair fight," said Benjamin, "anyway, the next day Hat wouldn't play or speak with Donald Gordon. In fact, although Donald Gordon and Hat went to Junior High School together, Hat just wouldn't speak to him, and in fact, he didn't speak to Donald Gordon until they were in the 12th Grade, some seven years later. They both were walking down the hall in High School, and both nodded to each other."

"You see, Hat had much respect for Donald because Donald whipped his ass. Donald Gordon applied maximum amount of power and energy into whipping Hat's ass and Hat never forgot that. America keeps making the same mistake when it comes to war; we need to really whip somebody's ass so they and everybody else will think twice about fucking with America, and we need to beat that ass good, like Donald Gordon did to Hat! So, when you shout "DONALD GORDON" you are confirming you understand your job requires you to give the enemy everything you have, never holding any force back, you shout "DONALD GORDON" when you apply maximum amount of fire power upon the objective. Maximum aggression, not a proportional response or measured proportional answer as is the United States of America way of responding when the enemy kills an American. Kill an American, oh let's bomb an airport runway!

aggression!" Benjamin and AB laughed, fist fived and yelled "DONALD GORDON!" The captain and the Interpreter drove with the Cook. The Cook was talking, and the Interpreter repeated to the captain what the Cook was saying. "Why are the Americans wanting to help me. I tried to get them killed because of my family. I have nothing to give to the Americans, I have no money, no riches, no fame. I am just a cook praying for forgiveness from Allah. I was wrong, but I had no choice. My wife was defiled." The Interpreter spoke before Captain Ward could reply "You and your family were tarnished by other Muslims. We kick down doors looking for Insurgents looking to kill Americans. Don't play innocent with me cook. You were paid well, to cook for us. We put food on your table." Just then Captain Ward spoke

up "I am not sure what you are telling him, but you need not explain our disposition." The cook chimes in to slow down the vehicle.

We dismounted about four homes from the intended target and set up flex security positions with only one back up position. The cook pointed out both his home and his brother's home, because he believed his brother's home had insurgents also.

Hat, Joey, Lonny, and I were going to enter the cook's house, while Benjamin, AB, Barnett, and Royster would enter the brother's home. The interpreter had taught us how to say, "Get down!" (NATHLAK) in Arabic, which should help the civilians. Once we were in place, Captain Ward gave us the word and Hat kicked the door down, yelling, "Get Down Get Down!" (NATHLAK – NATHLAK) While squeezing off three rounds as a male body fell to the floor. It was dark and I was behind Hat, and I couldn't see a thing, I just heard a big stump hit the floor. Captain Ward, the interpreter, and the cook were behind us, and then suddenly the cook hit the light switch. Two insurgents were

on the floor: one dead and the other playing dead. Lonny greeted the insurgent playing dead with the butt end of his weapon.

So, we counted two insurgents but the cook, said there were four, while yelling the names of his family as Joey checked the rest of the house. Bullets rang out down the hall as Joey sent four quick burners into the body of the last two insurgents, who had the wife and son

held up in the back room. With her clothes torn, the wife ran out to greet her husband, followed by his son. They hugged and kissed but the cook's little girl was nowhere to be seen. The cook, his wife and son were all calling her name out as they ran outside looking for her. The interpreter told Lonny to stand-up the insurgent who kissed the butt of Lonny's weapon, and asked him where they'd put the little girl, but the insurgent was just laughing. Suddenly, the cook stopped talking to his wife outside and ran back into the house and jumped on the back of the insurgent. Now the interpreter and the Iraqi Cook were both yelling at the insurgent in Arabic. I looked at Hat and saw that he had his trigger finger on his M4 and was slowly rotating his weapon towards the stomach of the laughing insurgent. Unexpectedly, the Cook broke free from Captain Ward and jumped on the

back of the insurgent again and in doing that, he bumped Hat and his arm swung out as a round was squeezed. Suddenly everything goes wrong. The Cook's little girl was being held by a fifth insurgent behind a false wall of sheet rock. She had been trying to break free from the insurgent who held her behind the false wall after she heard her father, mother, and brother call her name. The insurgent had a gag over her mouth, and she could not verbally respond. She had finally broken free and was running towards her father when the bullet hit her. As the little girl hit the floor, her mother screamed. Blood is everywhere. I shot the fifth insurgent while Hat dropped his weapon and began CPR, while yelling for a medic. He was moving frantically, as everyone else was stunned. The cook yelled, "No!" and

pushed Hat, but Hat just continued CPR, crying, ignoring the cook beating his back with his fist. Everything happened so fast; Hat gently gave the little girl mouth-to-mouth while the cook embraced his wife and son, all crying. Then, about a few minutes later with no life

in her body Captain Ward told Hat gently, "She's gone," and with one continuous motion, Hat stood up, pulled out his knife and gutted the only remaining insurgent, shouting, "Laugh now, bitch!"

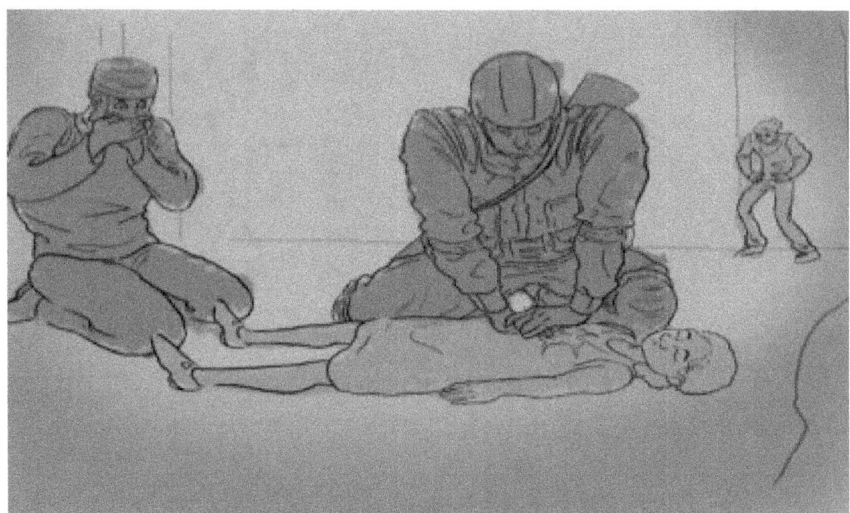

"NO!" the Captain and interpreter yelled but it was too late, as we all watched the body of the insurgent hit the floor. Now the Captain was giving first aid, and the medics who were called for the little girl arrived and were put to work on the insurgent. I could hear rear security warning us, "People are gathering in the streets," as the cook picked up his dead daughter and walked towards the door.

Hat walked towards the door too, and I hear the cook say, "ALI WILL; ALI WILL!" as his wife and son trailed him into the street to be met by 200 neighbors. I heard the captain saying, "Keep the insurgent alive!" AB reported all the insurgents in the brother's house were dead,

and that the insurgents had killed the cook's brother and his family. The streets continued to fill with people, as the captain escorted the medic with the still-alive insurgent to the vehicle and ordered everyone to mount up. Rear security called SSG Lewis and told him to leave base and get to our location quickly, as more people gathered in the streets. Hat was in a fog, walking towards the mob of people looking for the Cook, with Lonny calling Hat's name as he trailed after him. It was so intense that all the men were about ready to fire on the people in the street, as the captain yelled for everyone to stand down, and the interpreter ran to catch up to Hat. Now the entire squad was closing  in on the people in the street, Joey was screaming at the captain to allow the rest of the squad to open fire, while the Captain was yelling back, "Negative! Hold Fire!" Blackhawks were flying overhead, and AB was preparing to fire anyway when the Cook, with his daughter still in his arms, turned and walked towards Hat. Hat told him he was sorry, as the interpreter translated. The Cook said, "This is the will of ALI; it is not your fault. Go home; go home." Hat took the little girl out of the Cook's arms and held  her, and then a little old Iraqi lady took the little girl from Hat's arms and gave her back to the Cook. Hat repeated he was sorry and again the Cook said, "Go home." As they turn, Lonny screams at the crowd, "You knew these people were bad; why didn't stop them? Why didn't you do something about it? You knew! You knew!"

It was a scene I will never forget. We were encircled by these people, but I saw no hatred, just sadness. Hat, still soaked with the blood of the little Iraqi girl, was very quiet, as was the rest of the squad.

As we returned to the base, Cpt. Ward warned us of the likelihood of an investigation. When we entered the front gate, we were greeted by Command Sergeant Major (CSM) Ellington and the Chaplain. CSM Ellington said to Hat, "Son, things like this happen from time to time. Keep your squad moving; I need to talk with you." Sgt. Middleton and I stayed with Hat, but the Chaplain pulled Hat aside to pray with him. While the Chaplain was praying with Hat, the CSM gave us this speech about brotherhood and staying together like a band of unbreakable pack of wolves, like a band of brothers.

After the Chaplain stopped praying with Hat, Hat walked away from him and approached Sgt. Middleton saying "I was just told One was brutally beaten three days ago. He's dead; he's dead, just like that little girl."

I looked at my watch and the time was 23.45 hours. This was the worst day of our tour and it happened with 15 minutes remaining in the day in which Lonny proclaimed we would have a bad day. Lonny was right! Hat would be flying home 72 hours ahead of the rest of us, to attend the funeral of his cousin, son of his Aunt Lacy: The guy Hat called One, his brother.

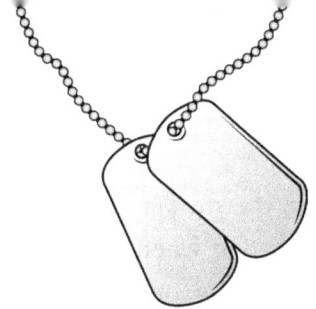

CHAPTER FIVE

After Hat informed Sgt. Middleton and I about the death of his brother One, I stayed with Hat as he changed out of his blood-soaked uniform and took a shower. He didn't say much. Sgt. Middleton did not say much either, disappearing after Hat went to the shower area. Joey, Benjamin, and AB took advantage of the time Hat was in the shower to brief CSM Ellington about the situation at the house of the Cook and the situation with Hat. Benjamin told the CSM that he thought Hat had Post-Traumatic Stress Disorder (PTSD). Joey and AB began to argue with Benjamin about his so-called medical diagnosis. Joey angrily said, "How the fuck do you know about PTSD, are you a fuckin doctor no, then shut the fuck up!" CSM Ellington calmly said "I guess the pack of wolves' thing didn't stick with you guys. Move then like a flock of flying birds, stay together, no matter what head wind you face." After hearing what the CSM said, AB grabbed Joey and Benjamin and they hugged it out. While Hat was in the shower, I tried to cheer him up with small talk, saying, "Just think, in three days we'll all be back at Ft. Bragg!" Hat just looked off into space.

A runner came into the shower area and informed Hat he needed to report to the TOC, and when Hat got their Captain Ward and Colonel Aponte, the Battalion Commander, were positioned behind a desk, waiting. Colonel Aponte said, "SSG Hadley, I am going give it to you straight. Your brother was killed four days ago. We're not sure why it took four days to receive this notification, but as you know, we are rotating back to Ft. Bragg in three days. I am sorry for your loss. We arranged for a commercial flight that would give you enough time to make the funeral, you will need to report to Ft. Bragg in ten days. Your flight leaves in eight hours. Our preliminary investigation indicates the shooting of the little girl was accidental, and the stabbing of the insurgent was due to combat distress. When you arrive at Ft. Bragg in ten days, you will report to the (TMC) Troop Medical Facility for a complete examination. How do you feel, Son, are you OK?"

Hat replied he was eager to attend the funeral, saluted and returned to the tent where he pulled out his dress blue uniform and with detail, placed all his awards on his uniform per the regulation. Sgt. Middleton, Joey, Benjamin, Lonny, and AB all returned to the tent together, and they seemed to act like nothing happened; they were jovial, telling Hat, "See you soon!"

Joey asked Hat, "Are you going to see Lynnette?" and Sgt. Middleton said, laughing, "Man, Lynnette is fine, trust me, Blue Nuts is going to see her and his son, isn't that right, Hat? I even know what hotel you'll be going too!"

Hat smiled at that point, so I quickly figured out what they were doing. SPC Barnett, the tent barber, gave Hat a nice haircut, while Joey continued to talk about Hat's girl Lynnette. "Yeah, I remember the first time I saw Lynnette, do you remember that Hat?" Joey said.

"Boys, we were in High School, and I was standing outside with my Italian brothers when I see the most beautiful black girl I have ever seen in my life. BADDABOOM BADDABING! She was walking along the fence towards the entrance of our high school; we were in the 10th grade. What I did NOT notice was Hat and Marty were 20 yards behind her. So, my crew of Italian brothers did the lean back DAMN THING from the Friday movie and out of nowhere Hat and Marty are up in our face, so I said to Hat, "Do you want to get your ass kicked? Get the fuck out my face! We're standing twelve legs and you have 4, you best back the fuck up!"

"Hat says quietly, "That's my girl."

"I said, "Congratulations! Shit! Damn! she is fine, so what, you wanna fight over something you already have? Check this out, your lady is fine, you know she is fine, therefore, I am going to look at her just like everyone else, if you don't want people looking at your girl you need to get an ugly girl! What happens if you were standing here and the roles were reversed, and she was white?"

"So, Hat looked at me and started laughing, then we all see a fine-ass Italian girl crossing the street, and we all say "DAMN!" - and we became friends for life that day."

Joey, Sgt. Middleton, and Hat fist fived and laughed, which was the first time Hat laughed in a long time. He told us that Lynnette would probably be at the funeral, and Sgt. Middleton told Hat to take it easy, because Lynnette would put it on him. Then Sgt. Middleton told Hat to call Rick when he landed; and not to go to the funeral home without Rick.

The Colonel's driver arrived right then to take Hat to the Airport. I wanted to suggest we pray, but fear took over my mouth and I didn't say a word. Lonny hugged Hat, grabbed his hands, and said, "Father GOD, gird this man up and point him in the way you want him to go!" Then we all joined hands together as Lonny continued "Keep his family safe and rest his mind, body, and soul. We are counting on you, Heavenly Father for traveling mercy; and traveling Grace Amen."

Hugs were handed out. Hat had to take a military flight from Iraq to Kuwait, and then he would catch a civilian flight to New York.

Hat's Story

The flight from Iraq to Kuwait was short, but my mind was racing, going a hundred miles an hour. By the time my mind unscrambled, I found myself on my civilian flight. I'm not sure how I got from one plane to another. The nice flight attendant on the flight to New York asked me if I needed a pillow, and I told her no, but obviously my face gave her a different answer as she returned with two pillows and a blanket. I had a window seat, so I placed both pillows against the window and threw the blanket over my head making my little space dark. Asleep, I began to reminisce about my cousin One.

I remembered when my Mother first told me to take care of Robert AKA One AKA 3 Wheel. Taking care of One was an easy task because he just followed me everywhere, I went. Whether playing with Marty, Rick or both, One was there. Marty called One '3 Wheel' and Rick called One by his real name, Robert.

One caused me to receive the worst whipping my mother ever gave me. One day, my mother and I were waiting for the number 2 train at the Rockaway train station. The Rockaway train station is an elevated outdoor Station, where the numbers 2, 3 and 7 trains run on the same track. We were on the platform when a train approached. It was the number 3 train, and when the doors opened a lot of people got off, so my mother held my hand tight and backed me up near the platform framed wall. From a distance I heard a man yelling, "HOLD THE TRAIN! HOLD THE TRAIN!" I moved up slightly to see who was yelling, but my mother squeezed my hand giving me a non-verbal signal to move back into the protected space she'd created for me.

I heard heavy breathing and thunder as this man was yelling for someone to hold the train as he climbed the platform steps. Just as he reached the last step to the platform, the train pulled away. The Heavy-set man dejected and with disgust took both his feet and jumped onto the door ledge of the train and rode along the door for one second then jumped off.

That man needed to make that train. That man also taught me how to train ride. The next day, Marty, One and I went to Rick's Fathers' dry-cleaning store to see if Rick could come out and play. We always contacted Rick using the back door of the dry-cleaning business, once we'd walked by the big front window to see if Rick was in the store. I don't think Rick's Father liked him playing with Marty and I because he never let Rick play when he was in the store. This day was no exception.

Rick was barely out the back door when his father grabbed him by the ear and took him back into the store. Rick's Mother was different; she would let him play with us. With Rick locked in and looking at us from the big glass front store window, we proceeded to the train station, sneaked under the turnstile, and headed up the stairs where we were stopped by Dirty Burt.

"Boys, where are you going? Where's ya mama? What ya doing?" he asked us, but I answered nothing to every question and pushed One up the stairs as the train approached. I told One to stay, as I pinned his back up against the platform wall.

"Stay here!" I repeated my words.

One said, "Are you leaving me? Are you getting on the train?"

"No!" I replied, "STAY HERE!" pinning him again as people from the stopped train walked by me. As the doors closed, I told Marty to watch me. I stood by the door when it closed and as the train started to move slowly,

I jumped on the doorway ledge. I rode the train until it picked up speed, and then I jumped off, running with the forward momentum of the train. When I turned around Marty was smiling and amazed; his jaw was wide open. As I approached him, he said, "You did it; you did it!" I said, "My feet were moving fast."

Marty said impatiently, "I'M NEXT!" and One said, "Me too!"

Marty and I both pushed One. Marty said, "3 Wheel you better not move from this spot!" and I pinned One's shoulders back against the platform wall and told him "NO!"

Marty and I waited with anticipation for the next train. As the train approached, Marty said, "Don't ride the train, mark where I jump off."

"OK," I said, as the train pulled into the station. After everyone exited the train and with the train doors closed, Marty dropped his hat at the point where he got on the train door ledge. I walked ahead about three car lengths, where I thought he would jump off. As the train slowly pulled away, Marty jumped on the door ledge. The train picked up speed, as Marty held on for dear life with his eyes getting bigger and with a smile on his face as he passed the position on the platform where I thought he would jump off at. After Marty passed my position, he jumped off running fast with the momentum of the train. When he turned around, he said, "You thought I was going to jump off at your spot, right?" smiling and laughing. Marty was excited that he'd held on longer than I gave him credit for. We were both full of smiles as we walked back to One, who told Marty that his hat was blown off the track. Marty looked over the edge but refused to get close enough to look for his hat. Still smiling, Marty said, "Let's do this again tomorrow. I plan on holding on longer than you, Hat."

As we walked down the stairs One said, "I wanna ride tomorrow!" "No, One," I told him, "You're too young, and you better do as I say."

At the bottom of the stairs stood Dirty Burt, he didn't say a word, so we just looked at him and walked by. As we crossed the street, we saw Marty's hat, which was all dirty due to cars running over it. Marty didn't care and picked his hat and put it in his pocket. As we walked by the dry cleaners, we saw Rick still looking out the big front glass window.

As we returned to the block, we told all the other kids about train riding, and the next day seven kids were waiting for Marty and me. We told them to meet us at the train station. We walked past the front window of Rick's Fathers' dry-cleaning business and Rick ran to the back door. His Mother, not his Father was working that day. Rick yelled at his mother, "Can I play with Hat today?" and she replied, "Sure," so Marty ran back to his house to get Rick a pair of sneakers.

Rick was Jewish, and his father had the only dry-cleaning business for miles. Rick never wore play clothes, so Marty would always give Rick play clothes to wear. As Marty ran to his house to get Rick some sneakers, I explained to Rick what we did yesterday and what we were doing today.

Once we reached the train station, I told the other kids to "sneak under the turnstile and meet me on the train platform." It was very difficult getting everyone on the platform, and as time passed Marty caught up to us and joined us on the train platform. He brought back some chalk with him, to mark the spots of the jump-offs. Rick was last sneaking under the turnstile, and when I asked him what took so long, he said, "I was scared."

I explained to everyone what they had to do, but Marty told everyone to just watch him. As the train approached everyone stood back, and once the door closed and the train started to slowly move, Marty jumped onto the train door ledge, rode the train, and jumped off. I marked his spot, and as he turned around all the kids were yelling, "OH MAN! OH MAN!!... I'M NEXT; I'M NEXT!"

I laughed and told them the train had enough doors for everybody, and when the next train arrived, everyone but One and I rode the train.

Rick was the first to jump off. I noticed that the doors seem to open for the most part in the same places. Marty started bragging so I told him to

get on the door ledge here and I would mark his jump off, and then I told him, "I am going to beat your mark." Everyone sized up another person, attempting to ride longer than each other. One even called out, "Rick!" so I had to put him in his place. When the next train arrived, I marked Marty's door then I went about four car lengths anticipating his jump off spot. Marty jumped off after my position on the platform again, his arms and legs were pumping. As I marked his spot, we both smiled and laughed. I then told him I'd beat that spot.

When the next train arrived, I gave One my 'you better stay' look. I rode the train past Marty's spot, and he told me, "OK, now I'M going to beat YOUR spot." We bet a grape soda. As the next train approached, Marty took his starting spot and I took a door in front of him, but as the train picked up speed, I heard a stomp as a body hit the platform. I turned and saw it was One.

I jumped right off, as my heart skipped a beat seeing One on the platform floor. Rick picked One off the platform and he was bleeding and crying, but the cries of, "Jump off!" by the other kids were louder than One's cry for help. Marty was still trying to beat my spot from before.

He did not jump off. I turned and started yelling at Marty to jump off, but the train had picked up speed and Marty was barely holding on. All the kids were yelling at Marty to jump off while running after the train, which was moving very fast. With nothing to grasp, Marty fell from the train door ledge and hit the walking train rails along the train track. When he hit the rail, he bounced up, then, came down. Marty was barely holding on as the gravity was calling him to fall 80 feet to his death.

By now I was screaming, "Hold on! HOLD ON, MARTY!" Swiftly, Dirty Burt was somehow standing 80 feet below Marty yelling the same thing. Dirty Burt grabbed a dumper that had a couch next to it, placed the heavy couch in the dumper, got into the dumper and shouted to Marty, "Let go! I got you, son!"

I made it to Marty just as he lost his grip.

His nose was bloody. Marty said, "I knew you weren't going to drop me!"
"I wasn't going to drop you," I replied.
Marty heaved a sigh and looked at me and said quietly, "You won."
When we got back to the platform, I ran over to One who was still sitting in the same spot where I'd left him with Rick. I picked up his two baby teeth from the platform floor, and carried him down the steps, with his knees and mouth bleeding. He didn't cry until he saw Dirty Burt at the bottom of the stairs.

Without a word, Dirty Burt took One from my arms and carried him home. Marty went home as well, with his nose still bleeding. On the way, Dirty Burt gave me a lecture on responsibility, but I was only thinking about the ass-whipping I was going to receive from my mother. As the elevator reached the 16th floor, I remembered my mother telling me to take care of Robert, and a big ass knot hit my throat. When Dirty Burt knocked on the door, my palms became sweaty and my mouth turned to cotton. I felt my mother looking through the peep hole and then I heard her scream as she opened the door. She had a full throat scream "OH MY GOD!" and of course One, who had stopped crying, is now crying again.

Dirty Burt entered our apartment and placed One gently on the couch, while my mother screamed, "What happened? You tell me right now what Happened!" directing her anger right at me while screaming all the way to and from the bathroom. She used soap and water to wash off One's legs and mouth, and then with no warning at all she swung at me! Dirty Burt, who was kneeling next to One, and slightly in the line of fire, just got up and walked out the door.

"You see what had happened was," I began, "we were on the train platform..."

"TRAIN PLATFORM?" She yelled at me, "GET TO YOUR ROOM AND TAKE OFF THOSE DAMN CLOTHES!"

I scrambled to my room, but my mother was right behind me breathing down the back of my neck. She shot past me, grabbed her belt and whipped my ass real good, all the while killing me with words like, "responsibility," and "you never listen."

That belt kept moving and my body kept feeling the pain, then after a while she stopped, and told me to take a shower and go to bed. That damn water hurt as it hit my body and the soap didn't help much either. It was still daylight outside, and I hadn't eaten a thing. I figured I'd rather go to bed with the sun still in the sky and hungry. My stomach was grabbing my attention. I thought briefly about asking my mother for some food but attracting my mothers' attention was not a good thing. My biggest fear right then was that she was still angry and would commence whipping my ass again.

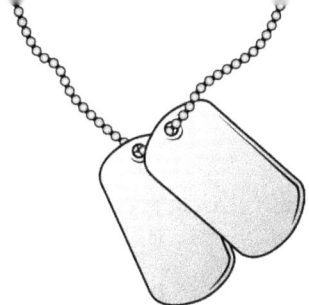

CHAPTER SIX

I kept an eye on One from that point on. Whether I was playing basketball or One playing Football, One always stayed close to me. Now One is dead just like the little Iraqi girl. Just like the little Iraqi girl. Just like the little Iraqi girl. That insurgent won't be smiling or eating without thinking about me; he set in motion something that will haunt me the rest of my life, but he will remember me for the rest of his sorry useless ass life.

When I landed at JFK in New York, I called my mother, who told me the funeral was in five hours. My mother also warned me that One had been beaten so badly that his casket would be closed. I thought, another closed casket just like Aunt Lacy's – One's mother. The funeral would be held at the Joe Morris Funeral Home, and she also told me Rick had called and said he would be running late and to wait for him before going to the funeral home.

"You better change the chip in your sidekick!" she warned, as Rick told her to say. I thought about Lynnette and Brice my son but, I knew I would see them at the funeral.

I thought about seeing Rick again; he'd gotten out of the Army after our second tour in Iraq. His Father was sick, and Rick had to run the dry-cleaning business, so he never re-enlisted. I figured that five hours before the funeral gave me enough time to get through the New York traffic, hit Sal's Pizzeria and pay a visit to the Shack record store.

As I walked through JFK airport making my way towards the Hertz Rental Car Center, a little girl accompanied by her mother said, "Look Mommy, a Soldier!" I stopped and shook the mother's hand before squatting down to ask the little girl (who now had a big smile on her face) her name.

"My name is Dorothy," she told me. I told her Dorothy was a pretty name, then I asked Dorothy's Mother if it was okay to give Dorothy something and she nodded agreement. I gave

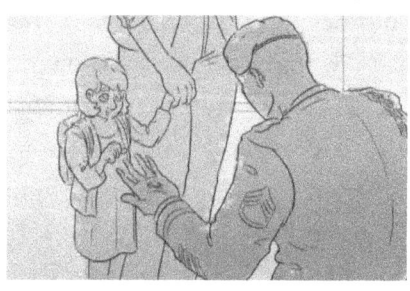

Dorothy the shiny airborne wings from my uniform, and the little girl smiled.

"What do you say?" the mother asked her, and Dorothy said," Thank you!" and gave me a quick hug. I said thank you back to Dorothy, stood up, said "Ma'am" to Dorothy's Mother and continued my way towards Hertz. As I walked towards Hertz, I noticed some folks smiled at me, while others just looked through me like I wasn't

there. I'd seen those same expressions in Iraq. I guess it doesn't matter what continent you visit, the hate look is worldwide.

When I arrived at the Hertz counter, I ordered a black convertible Mustang. As I drove off heading to Sal's and the Shack, I decided to turn on the radio and to my surprise I heard Toby Keith. A black man driving a black convertible mustang with the top down, singing a Country song his good friend Utah had taught him, wow what a sight!

Oh, and I don't want to die for you,
But if dying's asked of me,
I'll bear that cross with an honor,
'Cause freedom don't come free.
I'm an American soldier, an American,
Beside my brothers and my sisters I will proudly take a stand,
When liberty's in jeopardy I will always do what's right,

I'm out here on the front lines, so sleep in peace tonight.

I turned the music up as loud as the radio could go and sang at the top of my voice with Toby Keith as I drove down the highway, thinking of Utah and my squad. Man, whenever Marty and I would argue over who was the better rapper while we were lifting weights in the weight tent. It would get crazy. My favorite was LL Cool J and Marty's favorite was Jay Z. When we got to loud, Utah would always chime in with his Toby Keith, blasting and drowning our music and feud out.. When I woke up from my driving daydreaming, I found myself in the 'hood with another Country song blasting from the radio. Embarrassed in that environment, I found Hot 97.1 - Blazin Hip Hop and R&B. I parked the convertible with the top down in a bus stop across the street from Sal's Pizza and the Record Shack, which are conveniently next to each other. I needed to see the car from both windows of both stores, because if the right Police Officer was walking the beat, I wouldn't get a ticket.

After I parked, I ran across the street and ordered three slices, telling Sal "Let me have one now, and put the other two in separate bags." Then I went next door to the Shack. Tony the owner greeted me; "No

limit Soldier Boy; what's up?" as his only worker, Chris, gave me a fist five and smiled. While I was eating my pizza, I was pointing out the CD's I want that are hanging on the walls, while keeping a good eye on the Mustang that was illegally parked in a bus stop. Tony was very smart; he redesigned the Record Shack so that kids could download songs legally and illegally from the booths he constructed. He even gave them the websites they could search to download music. Tony charged ten dollars for 30 minutes per booth. Tony gave free booth time to kids needing to do school projects. These kids did not have WIFI in their apartments. When possible, Tony and Chris would help these kids with their school assignments. I always felt like Tony was some kind of super hero. He told me that Sal wanted to buy him out a few years ago. He offered Tony a great deal of money to expand the pizzeria. Tony often talked about selling the Record Shack and moving to Tampa Florida. As he contemplated Sal's offer, a young fourth grader came in to do an assignment that

required internet access. Tony knew at that moment; he could never sell to Sal or anyone else because of the project children.

Two kids walked in and rented a booth while I was there. He knew he could not make a good living by selling CD's. Tony, a Bush hater, usually gives me his 'I hate Bush' speech when I visit his store. "You know why you guys are taking back communities you cleared once before? One word – RATS! Rich people don't know how to deal with rats!"

With my mouth full of Pizza, I mumbled, "What?" "Listen," Tony said, "when you chase rats they run, and the insurgents ran to the border where the door was left wide open because FAT ASS RUMSFELD tried to fight a war on the cheap!" "Chris, where does he get this from?" I asked, and Chris replied, "The barber shop!" "Chris, close the door," Tony instructed, then he threw a wad of paper on the floor and took a broom and chased the paper around on the floor. "You see, with the door closed, the rat will run around in circles where I will catch it and…" he took the broom and smacked it hard on the paper, "THAT'S what you do! But no! Rumsfeld lets the bums run to the border, pick up friends and come back again. I tell you; rich white people are dumb. Yep, fuck Bush!"

I said, "Now you know, Tony, if you were in another country and you said that, about a President, you'd be imprisoned and or shot and hanged."

"Well, I'm not in another country," said Tony, "I am in America -where I can say what the fuck I want to say because Soldier boys like YOU! Keep it that way, so I say fuck Bush! Besides, someday, crazy white folk will believe in some stupid political motherfucka shit and will need a great black man to rescue them, like me. The second black president that knows how to deal with rats. And I definitely won't be a pussy and put military folk in harm's way and play political retaliation bulshit." And then Tony slaps Chris an high five.

Just then, a cop walked by, so I ran out of the Shack and stopped quickly at Sal's to quickly grab a slice of pizza in a bag before heading across the street towards the cop. His back was to me, and as he turned in my direction, our eyes met. Damn! He wasn't the Police Officer I hoped he would be. I smiled anyway and offered him a bag containing the slice of Pizza, but he slapped the bag out of my hand and the pizza hit the ground. He said, "that uniform you are wearing certifies, you can read, parking in a bus stop is a double fine." But just then Police Officer Nelson came from around the corner. He smiled as he walked towards the both of us, "Would'ya look at that uniform! Glowing after me, hey, where are your airborne wings?" Putting his hand on the shoulder of the other Police Officer, he told him, "Stop writing; I got this one." The other Police Officer, Jefferson from his name tag, was pissed and walked away saying, "Move this fuckin vehicle!" over his shoulder.

Police Officer Nelson hugged me, while Sal, who watched everything from his window, walked over with the other bag of pizza I ordered and gave it to P.O. Nelson saying, "Jefferson is too serious, Nelson; you should train him better."

"Thank you, Sal," said Nelson but he had to turn his attention to the bus driver who stopped and complained about my car to no avail.

"Keep it moving; keep it moving!" was all he told the bus driver, with a mouth full of pizza.

When he'd finished his pizza, he wiped his mouth and said, "I was sorry to hear about Robert, you know I coached him for basketball and football, he was disciplined; I don't know why he of all people joined a gang."

I said One wasn't in any gang, but Nelson said, "I saw the crime scene pictures; he died with his gang signs, Hard core, that kid."

"ROBERT WAS NOT IN A GANG!" I repeated, but just then Chris from the Shack ran out and gave me the CD's I had pointed out and said, "Tony said you can pay him later. I looked back at Tony in the doorway, and he waved.

As I got in my Mustang, I asked Nelson if I could see those crime scene pictures. He said, "Yeah, just come by the Precinct. By the way, where's your wing man? The bighead boy you are always with?"

I smiled and said," he's probably landing in Ft. Bragg right about now," then Nelson hit the rear end of the car with his hand and said,

"Now go! get your illegally parked ass out of here!" Shouting, "You're illegally parked!" as I drove off towards Joe Morris Funeral home.

As I pulled up to the funeral home, I saw about 15 teenagers in front of the place.

They looked too young and non-college-like to be One's friends I thought, as I drove slowly to a parking spot about three car lengths from the front of the funeral home. I returned the convertible top to its locked position  and got out of the Mustang, but as soon as I got out of the car I heard, "Shit that ain't no soldier boy! You ain't shit, HERO! Where's the cavalry punk bitch? You better step off or get your bitch ass beat to death just like your bitch ass cousin!" I attempted to keep walking, being as I was only three steps from the Funeral door, all the while, keeping my eyes on the young joker that was doing all the talking. Just as my arm reached for the funeral door, someone grabbed me and turned me around, so I reached out with my other arm and checked the chin of the young joker who was  talking. I was able to grab another kid real hard on his windpipe. Fists were flying at this point, as I was beaten and kicked to the ground, but I was able to check another chin of one guy before I covered up. I guess the noise

outside the funeral home caught the attention of someone, as suddenly everyone in the funeral home came running outside. A foot caught the back of my head and knocked me out. As I blackout, my eyes caught Timberland and Nike logos as I blacked out.

When I came to, I was on my mother's couch. My head and body were aching, but Lynnette's beauty quickly eased some of my pain. She kissed me gently as I looked around and saw my mother, Sister, and Rick.

"Rick pulled up at the same time we came out of the funeral home." Lynnette explained.

"Man, they looked like they were going to go after your mother!" Rick said, "Man, they all had guns – I'm not sure why they didn't shoot you; they were holding up their pants while they were stomping you! They looked stupid and you are crazy for walking into the lion's den like that, you're not bullet proof."

I attempted to get up, but Rick and Lynette stopped me, as I fell back onto the couch. "Stay here, baby," Lynnette said, as my mother gave me Tylenol and Lynnette held my hand.

I looked at Rick and he knew what I was thinking. He sat next to me and told me he'd booked my hotel room and had a plan, but we need to get rid of collateral damage. Then he went into the kitchen with my mother, which gave me a chance to tell Lynnette how I missed her and Brice.

Just then my mother came in from the kitchen where she must have been speaking with Rick, because she told me she wasn't going any place! I said, "Momma, please take Lynnette, Brice, and Lashae to our people in Santee, South Carolina!"

"I am not leaving!" cried Lynnette.

"I am not leaving either!" said my sister Lashae.

"Momma," I said, looking at them all, "I'm going to find out who killed Robert. That means you do NOT need to be here. Santee is safe, so please go."

My Mother looked me right back in the eye and said, "I did not raise you to die in these streets!"

"Momma, I'm not going to die. If anything, I should have died in Iraq. I need you to take Lashae and Brice to Santee (Eutawville), Lynnette will join you all in a couple of days."

Lynnette kind of smiled at my statement but also frowned. Rick took my mother back into the kitchen, so I asked Lynnette to help me up. I told her to pack an overnight bag, as I walked slowly to the kitchen. Rick gave me the SIM card for my cell phone, and I sat down next to my mother and told her I'd be back to take them to the airport. "Lynnette is a good Mother to your son," she said, and then it all came pouring out of her.

"Those boys downstairs look at her and try to talk to her, but she keeps right on moving, working, going to school at night. They offer her large amounts of money, but she knocks them down with words. Those boys are drug dealers' baby, they work for Leon and sell drugs in broad daylight. I'm not sure how they knew Robert, because he didn't stay here. He lived on campus and would visit me from time to time, but they found him in an uptown apartment on the Eastside dead with a girl who was also murdered. Baby, how could Robert afford that kind of apartment on the Eastside?"

She started to cry, "Robert's face was bashed in. The Police said he was brave to the end, displaying gang signs. They said he was a drug dealer, but he told me he worked at the bookstore on campus. I promised Lacy I would take care of Robert!" she sobbed loudly.

"Don't cry, Momma," I told her, kissing her on the forehead, "don't cry. Momma listen, "you MUST go to Eutawville, South Carolina, It's safe there. Momma listen, there are some questions you need to answer, like what happened the night Aunt Lacy passed? and who is my father?" but just then Lynnette walked in, holding a sleeping Brice.

I walked with Lynnette to the back room to place Brice in bed, and then I kissed him ever so softly. He woke up and said he needed to go to the bathroom, so I took him to the bathroom and when he finished, I took his hand to lead him back to his room, but he said, "I need to wash my hands first, Daddy!" I almost cried when he said that. I had been putting money in Lynnette's bank account for Brice since before he was born but she has never touched it. She's holding it for when Brice goes to college.

"I'll be back tomorrow," I told my mother, leaving with Rick and Lynnette as we headed to the hotel.

In the car, Rick told me he had everything under control and that he had a big surprise for me the next day. We arrived at the hotel and before driving off, Rick said, "Donald Gordon."

"Donald Gordon," I replied automatically, but I wasn't thinking about Donald Gordon, I was thinking about Lynnette. Lynnette is a strong black woman with a strong mind and strong will. Everything was about her and Brice and her and Brice only. I'd been looking forward to being alone with Lynnette, sore body, and all.

In the hotel room, Lynnette was gentle with me, kissing me softly on my face and wounds. She straddled me and slowly took off her top, but what I noticed more than anything was the soft mattress and the king size bed. I then noticed her smiling as I ran my hands along her breasts, grabbing her shoulders and slowly twisted her on the bed as we both pulled our pants off. Lynnette had to help me with my pants as I struggled to bend over. Lynnette gave me a condom and entered the mother of my Son.

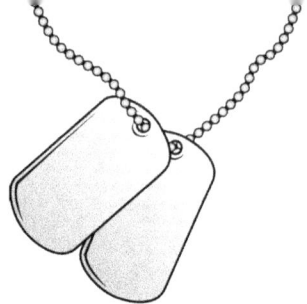

CHAPTER SEVEN

After making love and with Lynnette still in my arms, I told her "I love you" and she responded, "I love you more." Things then turned serious, as I said, "If you love me, why won't you marry me, Lynnette?"

Lynnette replied, "We talked about this before Hat; I just refuse to marry a military man. I want a man totally dedicated to his family; a man committed to ME only, not to a country!"

"I can do both," I replied, "but for some reason you're not willing." When I pushed her further, she finally told me something that surprised me.

"You know the man you call Dirty Burt?"

"Yes," I replied.

"Well, he was my father."

I hadn't seen that coming. Dirty Burt's body was found the day after he called Leon a punk bitch drug-dealing chump who was a crybaby as a little kid right in front of everyone. This happened about six years ago. Leon surprisingly just smiled and never said a word that day, but I guess actions speak louder than words, because Dirty Burt was brutally beaten to death the following day. I wiped the tears from Lynnette's cheek as she told me how embarrassed she was of her Father.

Lynnette told me that the war 'did something' to her father, because he just went crazy one day after returning from the Vietnam war, and so her mother threw him out. He received a check every month and placed the check under their door on the 1st of each month like clockwork. "I know what my mother went through, and I am not going to live like that!" she cried, "I LOVE YOU HAT!"

I shushed her and eventually we fell softly to sleep. In the morning, we made love again. During breakfast, Rick called and said to meet him at 11 am at our favorite candy store in the neighborhood, so Lynnette and I took a cab, and arrived at the candy store just before 11 am. That gave me a chance to buy a Nephi cream soda and split a bag of Wise potato chips with Lynnette before Rick arrived. Just as I was paying for the goodies, and with Lynnette standing in the store doorway and she called out, "Hat, you are not going to believe this, there's a big red truck with a white boy in the truck blasting Country music!" I smiled and quickly paid for the goodies and walked out of the store holding Lynnette's hand as I heard a familiar song coming out of a red Ford F250 with big tires: Another Toby Keith Song

> *'Cause justice is the one thing you'll always find,*
> *You gotta saddle up your boys*
> *You gotta draw a hard line,*
> *When the gun smoke settles we'll sing a victory too*
> *and we'll all meet back at the local saloon!*
> *We'll raise up our glasses against evil forces*
> *Sayin' whiskey for my men and beer for my horses, Whiskey for my men, beer for my horses.*

Man, two white boys playing loud Toby Keith Country music in the heart of Brooklyn! Utah powered down his window and yelled, "WHAT'S UP, NIGGA!"

I grabbed Lynnette's hand tighter as we marched towards the truck that had a big sign in the back window that said 'DON'T TOUCH THIS'. Rick got out and opened the door for Lynnette as I smiled up at Utah.

John Smith AKA Utah, from… you guessed it, Utah. Utah's Father did something so terrible within the Mormon community in Utah that Utah left and joined the Army. I met Utah in basic training; we were in the same squad, standing in a single line waiting to get our assigned M16 weapons. I was standing right behind Marty, and Utah stood two people behind me. The Drill Sergeants were screaming instructions on how to sign for our assigned weapons, but I took that time to talk to Marty when the Drill Sergeants walked by and were out of range from my voice.

For some reason, Utah took exception to me talking, and told me to "shut the fuck up." I turned and in a low voice told him to "shut the fuck up!" The guys between us moved, creating some air and opportunity for me to nose up to Utah, and right on cue Utah said, "What ya gonna do, Nigger?" I punched Utah on his forehead, but I was aiming for his nose. Utah backed up then he charged me. The other Soldiers were yelling "Fight!"

The Drill Sergeants arrived at the same time as Utah drove me into the pavement, and one Drill Sergeant said to the other, "How long?"

"One minute," the second Drill Sergeant replied. In the meantime, Utah was applying this grappling move on me. The Drill Sergeants pulled Utah off me, much to my delight because he seemed to be trying to fold me into a pretzel. Now three Drill Sergeants were yelling at Utah and I, while the other Drill Sergeants attended to the rest of the platoon and handed out weapons. After the three Drill

Sergeants finished putting on a screaming contest, they decided to make an example of Utah and me. After making us do face

to face push-ups, during which Utah attempted to do more push-ups than I, thus making it a competition, we were then told to get on our backs and do flutter kicks. Again, Utah turned that into a competition as well. Next,  we had to do the overhead claps, followed by the side straddle hop. By now Utah was tired of competing, and I was tired of the Drill Sergeants and exercising. I guess the Drill Sergeants must have been tired of screaming also, because they told us to stop, and then hold hands.

I reached for Utah's hands, but he pulled away saying, "I ain't holding that nigger's hand!" I was tired as hell, but somehow, I found enough strength to strike Utah in the jaw. He must have been tired as well because he didn't even try to duck my punch, but he did find enough strength to hit me back, right in the eye. We faced off squaring up, and Utah spit in my face. I hit him in the stomach and as he bent forward, I slammed the back of his head with my forearm and spit back onto the back of his head as he fell to the ground.

Just as Utah hit the floor and before I could stomp him with my boots, the Drill Sergeants stopped me in mid-motion. The Senior Drill Sergeant was now present, and smiled at us both and said, "Good! Unresolved conflict born of ignorance; I love it. Well, ignorance, move over and say hello to survival!"

For some strange reason I spoke up and said, "Senior Drill, what does that mean?"

"What that means, Soldier, is I'm your worst nightmare and I am going to burn your house down." I still wasn't sure what that meant but he had a strange look in his eyes, so I shut up. When he walked away, the Drill Sergeants started in on us again.

After the Drill Sergeants finished with their yelling competition, we were instructed to do eight count push-ups and sit-ups, then they made us low crawl. While we were low crawling the Drill Sergeants were yelling, "Go on, fight now! Fight now!" After low crawling we had to do squat thrusts, and by now my body had gone beyond muscle failure. After the squat thrusts they gave us a water break. We ran to the water fountain, and right about that time, my mind shifted from hating Utah to hating the Drill Sergeants.

I told Utah, "I got an idea; let's end this beef and pick this thing up after we graduate from basic training. We can go off base and fight later."

"Good," whispered Utah, "because I hate your black ass!"

The problem was, the water fountain was outside the window of the Senior Drill Sergeant's office, and of course with his window up and his window screen down, he heard everything. The Senior Drill Sergeant lifted his window screen and said, "to err is human, but in the Army sometimes it's costly," and then he called the other Drill Sergeants to his window. One Drill Sergeant told us to hold hands again, and this time, I reached for Utah's hand, and he reluctantly held my hand. The Drill Sergeant made us go take a shower together. They tied our bunks together as well, and then the Drill Sergeant escorted us to the Dining facility where we sat together.

I took some time to look around and spotted Marty, who had that 'I pity you' look on his face'. Everyone seemed to be looking at us. After a quick lunch, we had to put on our full combat gear, including a backpack that weighed 70 pounds. Once on the road march, Utah decided to walk faster than I, so the Drill Sergeant trailing us in a pick-up truck told me to keep up with him. So, I decided to pick up a jog and pass Utah. Then the Drill Sergeant told Utah to keep up with me. Utah ran past me, and I ran past him.

Finally, Utah said, "OK. Stop running," so I stopped running and he bent over and said, "you think you're so special. Your ancestors fought and died because of the word nigger, now you use it as a badge of courage, a right. You say, you say the word nigger to take away the power of the word but that's bullshit! You call it slang; you ain't shit! Fuck you nigga!"

I hit Utah in the stomach, and he fell to the ground, and said, "Nigga!" I punch him all over, and still he said "Nigga!"

The Drill Sergeant pulled me off Utah then poured water on the pair of us, telling us to get up and keep it moving. That damn water weighed us down now, so we were walking instead of running, we were so tired and bloody. When we reached the ten-mile marker, the Drill Sergeant gave us Gatorade and a Snickers bar. Utah and I sat and looked at each other, neither of us saying a word. I was thinking about how I was going to do the ten miles back.

The Drill Sergeant told us to get in the back of the pick-up truck and yelled, "Next time, we're going to bring you back out here at night. Then we'll see how you guys handle these wild Boars and Snakes."

When we got back, they made us take another shower together. Utah and I slept, ate, and trained together from that day on. Harmony ran throughout our basic training platoon. Nobody wanted to be in our shoes!

As Rick introduced Lynnette to Utah, I leaned forward to see how Utah was operating the gas pedals with his prosthetic legs. After the introductions, Utah said, "Let's go to Sal's; I want some of that famous pizza you talked about."

"Utah, man, what are you doing here?" I asked him once we were in the truck.

"What?" Utah asked me, "So I can't come and check you out? Man, we've got a game plan!"

I asked him what game plan, as Rick told Utah to turn left at the third light. Then Rick said "Remember those days Hat talked about Sal's? Well get ready, brother."

We couldn't find a parking spot on the street, so I told Utah to park at the bus stop right across from the record Shack and Sal's. As we crossed the street, I noticed that Utah walked kind of normal for a guy who'd had his legs blown off. He walked right into Sal's just like he'd been there before.

"Yo, Sal!" Utah yelled, "Give me two slices; no, make that a large New York Pizza." Sal turned around and told him, "We only sell New York Pizza in Sal's. Where you from Chicago or somefin?" in his heavy Italian New York accent.

Utah said, "I'm from Utah, Sal." Sal being the character he was said, "Listen PIEZON, why do you keep calling' me Sal? You know me or something?"

"Everybody knows the famous Sal's!" Utah told him, as Lynnette, Rick and I laughed under our breath. "I heard all about Sal's famous pizza in Iraq, and I heard about Sal's from some guys in Kuwait! Man, Sal, you're famous across the WORLD!"

Sal smiled, winked at me and told Utah the pizza pie would be ready in 12 minutes, and that he'd make a special pizza pie for him.

I grabbed Utah by the shoulder, laughing and lead him out of Sal's and into the record Shack. As we walked in, Tony was saying to Chris, "Rumsfeld and Wolferwitz are the real weapons of mass destruction, sending people to war without the right support. Then he turns and said "Hey! Soldier boy, pretty girl, and two white boys; four the hard way. Who you got there?"

Rick told him, "This is our friend Utah," and Utah said, "Nice to meet you, Tony. Heard a lot about you, you have any Toby Keith?"

"Utah, is it?" Chris said, "We sell hip hop and R&B," but Tony interrupted him.

"Like I was saying, Chris, and the no limit soldier here can attest to it, a black man and a 'nese tell the President, "And Tony tried to put on a proper voice here, "Well Mr. President, Rumsfeld and Wolferwitz are wrong. We need to send more than 160,000 military members into this war; we need to secure Iraq's borders, so insurgents can't get in or out of Iraq." Then Tony dropped back into his regular voice, "But who did the bonehead listen to? The two suits, that's who, instead of listening to the Stars. The Stars know how to execute a war; they attend war colleges!"

"What Stars", Chris asked him.

Tony said "I'm talking about Powell and Shinseki; man, they even made Shinseki retire after he went before congress and told them the truth, and you know how dumb Congress is! Why do we keep these dumb asses in Congress anyway? They always use the word fight. Them motherfuckers ain't never had a fight in their entire lives! They use fighting words, but those words are hollow like a motherfuckin water hose. I'll fight for the American people the Congress Men would say.

Translation – If I do at least five years, I qualify for a pension. Damn Soldier Boy gotta do at least 20 years!"

Chris interrupts "I thought you said 'nese?" Tony replied "I did say 'nese, fool! Powell is black and Shinseki with his smart ass is like Japanese, Chinese, Philippines - some shit like that. He's smart; he's Asian and a real military man. My point is dumb ass people always listen to dumb ass people. Powell and Shinseki are military men not regular Soldiers but sure-fire spit and polish men with only one goal in mind, doing the right thing for the Men and Women in uniform serving and defending our country and yet the President listened to self-serving assholes, OK? For instance, General Patton…"

"Hold up, Tony," I cut in with, "our pizza is ready. We'll be back!" and I quickly ushered everyone out of the shack."

"How does he know so much?" Utah asked me.

I explained to Utah "that on top of the counter Tony has hip hop magazines, but under the counter he has the Wall Street Journal, the New York Times, the Weekly Standard and a 357 Magnum Revolver."

As we walked into Sal's, Sal tells Utah he made him some Zepolees at no cost. I guess Sal must have noticed how Utah walked. As we dug into Sal's famous Pizza pie, Rick took a call from Marty, saying everything was set up.

Utah commented to Lynnette how beautiful she was in person, and that her pictures didn't really do her any justice. Lynnette laughed in surprise and asked him, "When did you see my pictures?"

"In Iraq," he told her, and Lynnette smiled and looked at me, and I smiled right back at her. Just then, you could hear Tony's mouth as he entered Sal's, "The hell with you, Sal!" he called out as he made his way to our table.

"Here, Tin Man," Tony said, "I made a CD of Toby Keith's greatest hits for you."

"Why did you call him Tin Man?" Rick asked Tony.

"I know Government legs when I see them." Tony replied.

"Damn!" I shouted, "More Toby Keith!"

Then Sal said loudly, "Boys you better move your truck; a tow truck is outside!"

I grabbed the pizza box with the left-over pizza and ran out the door, Rick slapped twenty dollars on the counter while I yelled at the tow truck driver "don't hook up." Utah is right behind me with Lynnette holding his hand, but Rick ran straight pass Utah and slapped the keys from him while barely missing getting hit by traffic. Rick then jumped into Utah's truck and pulled up about 25 feet past the tow truck. Of course, the tow truck driver was mad and yelling, but by now Lynnette was in the truck with Utah, and Rick opened the door for me and as I jumped in as we all laughed.

Rick drove off, and then turned down a street with a lot of abandoned buildings. As he slowly drove up the street, I saw three black Suburban's parked in front of the only standing building on this block. I asked Rick, "What is all this?" as we stopped at the Suburban's. Rick hopped right  out of Utah's truck, and yelled, "Donald Gordon!" and Marty, JB, Maurisee, Frost, Lonny, Benjamin, AB, Barnett, Cousins, Royster, Joey, and a new guy Garries came out of the SUV's. Garries worked with JB and Maurisee.

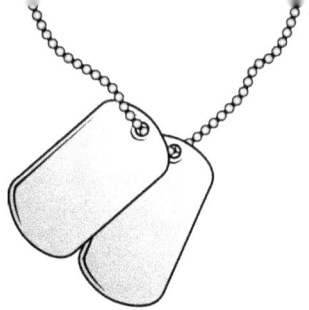

CHAPTER EIGHT

I couldn't believe all the squad was present. It was like a smiley birthday party; grown men were exchanging hugs all over the place as I introduced Lynnette to everyone. I did not know Garries so JB and Maurisee introduced their apprentice. JB said, "Garries is smarter with electronics and can pull apart any component and put it right back in its original frame." I said, patting Garries on the back and looking at

JB "Good to see you brother." I'd never thought I would see JB and Maurisee again, so man, it felt good! As we walked into this abandoned building, I noticed it had electricity, chairs, and tables. In fact, the chairs in one room were set up in two rows, and there were all kinds of equipment on the tables.

Marty told everyone to sit down, as Rick stood behind a photo projector preparing for a briefing.

Frost's Story

It was nice seeing Hat again, especially in better surroundings. The last time I saw him he had the blood of a young Iraqi girl on him; this time Hat was draped by a beautiful woman, clothed with Love.

After everyone was seated, Maurisee took Hat and Lynnette outside and showed them the Fortress alarm system. Maurisee, Garries and JB had placed computer-chipped ground sensor radars with advanced prism technology 25 meters around the circumference from the building. Remote controlled mini-guns and seven mini-cameras were placed on the roof and

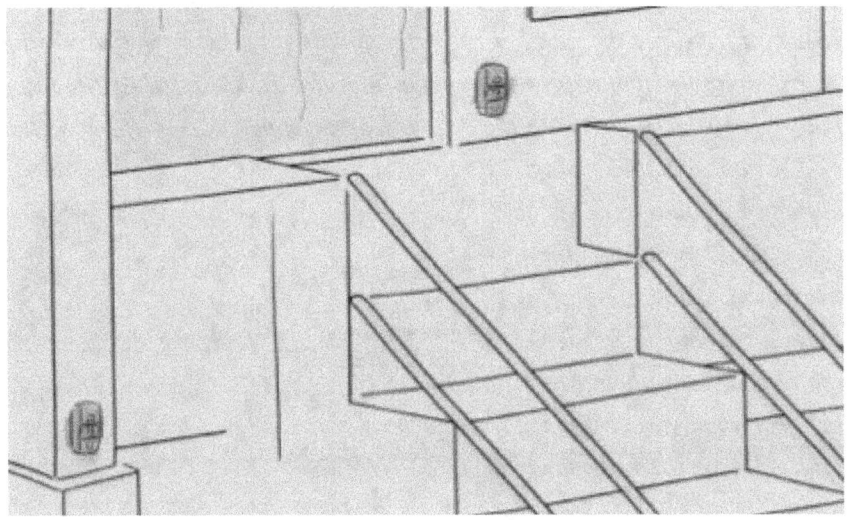

attached to the building, which also had flood lights. The remote guns were controlled by the brain of the computer, and were voice activated. The advanced prism infra-red technology identified intrusion in less than a second. The mini cameras were the size of a dime and along with the flood lights were all connected to the system, which was entirely powered through solar.

When a single infra-red beam was broken, the mini-cameras, floodlights, and guns were automatically activated. The computer recorded everything and sounded a voice-activated alarm. The system could also send an email alarm if you were away from the fortress.

"Hence, our modern-day fortress. Our sleeping quarters are secure and safe," Maurisee said.

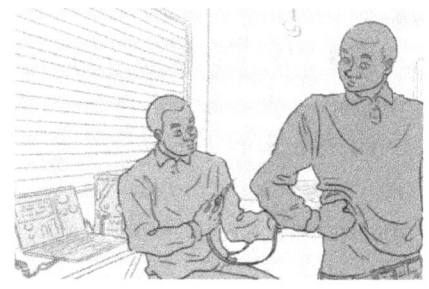

When they walked back into the fortress, JB was at the table on the right, giving Lonny an injection for the Blue Force tracking device invented by JB. JB injected Lonny right in his butt cheek. "Now we can track your every move," JB told him. JB was affected by the number of stories told about Soldiers going missing and their whereabouts being unknown in Iraq. JB simply used the technology for vehicles and with the help of Garries invented something better.

Maurisee called Marty over and hooked him up to this machine with his 9mm. Marty placed his shooting hand in a biometric device that would only allow Marty to shoot that gun. Maurisee said, "This is the best way to control bullets and create the ultimate gun control. No one can Fire this 9mm but you, Marty; it will misfire for everyone else."

Hat asked JB to inject Lynnette with the Blue Force tracking device, telling her he wanted her tracked, and Lynnette said she didn't mind getting the shot.

Once everyone had their weapon biometrically hooked up and had received their Blue Force tracking injection (along with the stupid designer Band-Aid Maurisee insisted go over the injection spot), Rick who stood impatient started his briefing.

"This is the enemy," as he put up pictures of 20 kids, none older than 13. "These 20 kids are security for Leon's operation, he's the drug king," and Rick took photos of Leon and placed them on top of the blackboard and on two walls. "These 20 kids do nothing but stand around with their cell phones, prepared to hit the redial button in the event the scooter kids give the alarm stating the Police arrived. The redial number is connected to one of the captains. The scooter kids ride around on scooters running drugs from corner to corner and are the first line of communication for intruders like the Police and strangers."

Then Rick put up pictures of the scooter kids - there were 10 of them. "The 30 youths rotate on three eight-hour shifts."

"They're all children!" Benjamin commented.

Rick ignored him and continued, "These six teens, aged between 15 and 17 are lieutenants. They in turn work for three captains. The first Captain is Poochie, a 20-year-old who likes to show off his gun." Rick switched from photo to video. "He keeps his gun - a 9mm - in his right pocket but he's left-handed."

AB scoffed, "He's stupid! He must cross his body to pull his gun out of his pocket!"

"You're right," said Rick, "he keeps the money he is supposed to collect in his left Pocket; Poochie pulls his bundle of money out of his pocket at least once an hour, counting it openly in public. Notice how he runs holding his pants; his money and gun weigh heavy in his pocket plus the dumb ass wears pants that are too big, and he has no belt on. He spends a lot of time talking to girls. Check this out…" as Rick shows a video of Lynnette cursing Poochie out because he touched her on the arm when she walked by him.

"I hate to be touched," she chimed in, "and his breath stinks with all that gold in his mouth!" Everyone laughed except Hat.

Rick continued. "The second Captain is Larry; he's Leon's cousin, he is right-handed, and carries his Glock in his right pocket. He also wears droopy pants, but unlike Poochie, Larry doesn't wear dreads but sports a small afro. The weakest captain is called Cross-Eyed Chris. He wears a patch over his left eye, and I've heard it's because he came up short on his money one day and Leon had him held down, then took a cigarette and burned his eyeball. Cross–Eyed Chris is the weakest link, and we should gather information from him."

"Wait," said Joey suddenly, "can you give me a close up of the person standing on the porch next to Cross Eyes?"

Rick zoomed in, and Joey said, "We have a problem. That's one of Mannatorres' guys."

"Who is Mannatorres?" Royster asked, and Rick said, "Yes, this guy doesn't even have a gun; all he does is stand on the porch and talk on the phone. He never walks more than two feet from that door, and I think the drugs and money are inside that door. He arrives around noon and leaves at 8 PM. Half of the time he sits in his car."

Lynnette nodded, "Yes, he's there every day; he doesn't speak to anyone."

"Mannatorres," Joey told us, "Is the drug lord of Canarsie. But why would he have one of his guys in the 'hood unless he is controlling the drug situation here? Man, this is bad news."

"Okay," spoke up Lonny, "we should go after Mannatorres AND Leon; let's go after the big fish!"

Marty cut that short. "No! The plan is to take down the complete running of this organization; capturing Leon and Mannatorres will not solve anything."

Hat spoke for the first time, "Listen, my plan is to find out who killed my brother, and then I am going to kill that individual, plain and simple.

We're not running an operation here; this is about killing, so wrap your mind around that. Whoever killed my cousin One will have his blood spilled."

AB wasn't happy, "Wait, I thought we were going to use rubber bullets… and… man… these are Americans! We can't go around killing innocent civilians!"

"Innocent civilians?" Hat told him, "These are low-life useless people who prey on their own for their benefit of money! Innocent? Man, these drug dealers aren't even American! They are evil devils who don't really give a shit whether they live or die, so I plan on helping them reach their goal quickly, besides are you saying that we only killed combatants in Iraq?"

"What I'm saying," AB replied, "we live by core values, Loyalty (Joey stands), Duty (Royster stands) Respect (Barnett stands) Selfless Service (Benjamin stands) Honor (Cousins stands), Integrity, and Personal Courage! Law enforcement is designed to take care of this problem. We're soldiers, not social workers, or lawmen. We support and defend the Constitution of the United States. If we do this, we are no better than the drug dealer!"

"Man!" Lonny shouted interrupting AB, "I did not come here to shoot somebody with rubber bullets! I heard Hat was jumped by 15 of these assholes! War is war, so I DECLARE WAR on these domestic terrorists!"

AB cut in, "You cannot call them domestic terrorists. You're giving them this arbitrary label, domestic terrorists so they will fit in some wrong box just to justify killing them! This is bullshit! and this is wrong! we live by a code."

Joey said, thoughtfully, "If Mannatorres is involved, we might be biting off more than we can chew," but Lonny said "Man, we hunted and killed Saddam, Mannatorres can catch a hot one also!"

"Look," said Marty, "Lets grab one of these fools like Rick said, and see what information we can retrieve, then we can make a decision."

I saw Lynnette touch Hat's arm and she slowly placed her hands in his. Hat responded to Marty's request, "Fine." I suspect Lynnette's touch had something to do with his response.

Hat, Lonny, Lynnette, Utah, and I took one of the Suburban's to pick up Hat's Mother and Son and take them to the airport, while Rick, Royster and Barnett took Utah's truck to pick up Hat's rental car and return it to the airport. They also picked up another rental vehicle. They were going to use the rental vehicle for a snatch and grab of Cross–Eyed Chris. Utah told them to make sure they cut off the car alarm, because if they didn't, they would receive the shock of their lives. Marty, Maurisee, Benjamin, AB, Garries, Cousins, JB, and Joey stayed at the fortress going over contingency game plans.

They knew things were going to get out of hand, and out of hand very quickly if they didn't come up with a plan to solve the murder of Hat's brother and take down the drug organization prior to Hat and Lonny taking matters into their own hands. They anticipated the squad would be split; half willing to kill for Hat, and the others siding with AB. Marty knew he was in the middle of loyalty, friendship and doing what was right.

While in the truck, Hat told Lonny to "be cool, we are going into the projects to get my Mother and Son, and the drug dealers will be around." Hat tap-coded Rick to find out at what level of command would be on duty right now, and Rick told Hat that Mannatorres' guy would be on the porch, and the lower-level Lieutenants would be out - but not the captains.

We parked at the candy store and walked across the street into the projects, and the scooter guys signaled our arrival. And, Just as Rick said, Mannatorres' man was there on the porch, together with some of the guy's Rick had identified. We made it to Hat's Mother's apartment with no problem, and his mother was waiting and packed. Hat's son Brice was extremely happy to see his mother and father.

Back at the fortress, Maurisee showed everyone the animated pictures of the Blue Force tracking. As we exited the building, one of the guys who jumped Hat at the funeral home told Hat he was surprised Hat was walking with the ass-whipping he'd received, but Hat instructed everyone to "Keep walking."

"Hat give me permission to shoot the bastard?" Lonny asked him, "This ass hole has no clue on how to Fire his weapon; it's nothing more than a decoration to him. He only knows how to hold a gun; he probably would miss a 25-meter target!"

"No," Hat said, "just keep walking."

I drove to the airport and watched the tearful goodbyes of Lynnette, Hat, and Brice. Hat told his Mother and Son he would call them and that their cousin, little Kevin Worrelly would pick them up from the airport. As we pulled away from the airport, Hat, with watery eyes received a tap-coded message from Rick. Rick told Hat they'd picked up Cross-Eyed Chris coming out of the supermarket and that they were holding him at the fortress.

Unbeknown to us, Rick was followed, after his grab-and-go in broad daylight. When we arrived at the fortress, Marty had Cross-Eyed Chris' legs and arms tied to a chair in a dark room with his mouth gagged. JB had Cross-Eyed Chris' cell phone and he was downloading all the information from the cell phone and ghosting the phone so that every time Cross-Eyed Chris would send or receive any communication from his cell phone, JB could monitor everything. JB also said he could use Cross-Eyed Chris' phone to correlate his location, by pinging the towers used by his cell phone to identify his position to within 100 meters. After getting that information from JB we huddled and devised a misinformation/psychology warfare game plan.

Lonny approached the gagged kid and noticed he had a 'No Snitching' t-shirt on. Lonny says, "No snitching? No shit! You have got to be the most ignorant person in the world! How do you spell snitching? No! Don't look down at your t-shirt you dumb mothafucka! Frost take the gag off his mouth!"

As I removed the gag, Cross-Eyed Chris began to scream for help, at the top of his lungs. Lonny let him scream a while then slapped the shit out of him like he did that cook in Iraq, saying, "Here, taste your blood!"

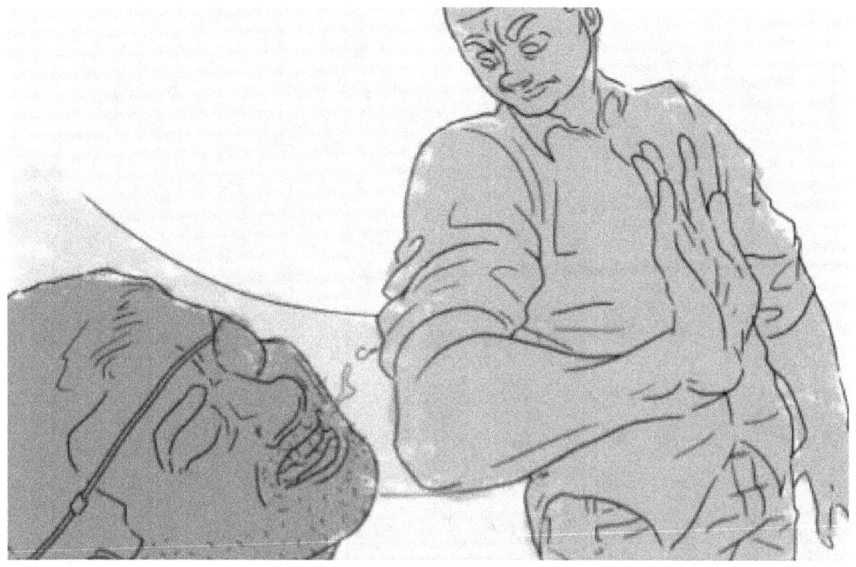

Cross-Eyed stopped screaming then, so Lonny said, "Well I guess you aren't that dumb, but wait…" then Lonny slapped him again and said, "Buy a belt, stupid! You look totally fucked-up running around holding your pants up. You're supposed to be a badass tough guy; so how did you get here in broad daylight? You're just a dumb ass!" Then Lonny slapped him again and said, "That's for when you are back with your boys', and you tell them how lucky I was because if you had the chance you would have fucked me up."

Then Marty walked in and interrupted Lonny's fun, "Now I'm going to ask you a couple of questions and if you give me the right answers you will go free, but if you bullshit us, well, your gonna have some major fuckin' problems for the rest of your shitty life." Cross-Eyed Chris said, "I ain't tellin

you shit! I ain't no snitch!" "Wrong!" Lonny said, slapping Cross-Eyed Chris in the face again, the same way he slapped the Iraqi cook.

Then Maurisee walked in with his laptop and told Cross-Eyed to look at the monitor. Maurisee showed Cross-Eyed Chris an animation of a leg being placed into a bucket. After the bucket bubbled over with smoke, the leg was

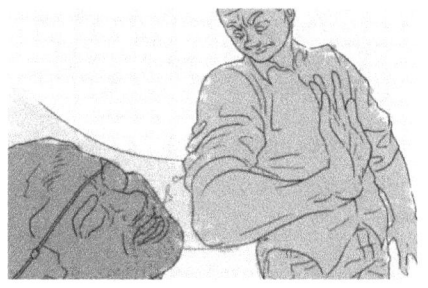

removed, and all the meat was off the bone. Then Hat brought in a bucket with smoke coming out of it.

"Are you talking?" asked Lonny, as he removed the ropes from Cross-Eye'd Chris left leg. Cross-Eyed Chris started to wiggle, attempting to get out of the chair.

"So, this is what is going to happen," Hat told him, "We are going to place your leg in this bucket, and you will feel pain right  away as the acid peels the skin and flesh and leaves only your bone remaining. But because you don't have a Government job, instead of your leg looking like this..." and Utah walked in and lifted his pants leg up showing his prosthetic leg, Lonny broke off the leg of a wooden chair and handed that to

Hat, who continued, "the State will give you this; a wooden leg! You'll be walking around with a patch over you left eye and you will walk with a wooden peg like it's a bicycle kick stand."

Lonny slid the bucket close to Hat, who said, "Now…" but Marty interrupted, saying, "Wait, he has a bad left eye; shouldn't we acid his right leg, to give him balance? I mean after all, he's holding his tongue for a guy who burnt his eyeball, so we should at least help this dude!" Then Lonny nudged the bucket closer to Cross-Eye'd Chris right leg.

"Wait!" Cross-Eyed Chris squawked, "Wait; what d'you want to know?"

"Hell no! No snitching! Let me put his leg in the bucket" Lonny said, but Cross-Eyed Chris said again, "No! What do you want to know?"

"I know," Marty said, "Let's do the right leg!" and Lonny reached down and touched Cross-Eyed Chris' right leg, and Chris started crying, "What do you want to know?"

"Who killed my brother?" asked Hat.

Chris said, "I don't know; the word was he owed Leon some money because he came up short!"

"Came up short how?" Hat asked.

"On a package. He was dealing, but only uptown" Chris replied.

"My brother was NOT A DRUG DEALER!" roared Hat.

"When does Leon come around?" Rick asked Chris, who replied, "I ain't telling you shit, you dirty muthafucka!" Then things got ugly again. Lonny slapped him in the face and Chris said, "Leon hits us on our cellies and we get the streets cleaned up before he rolls in."

The questions were coming from all sides now. "What about the guy on the porch?" asked Joey?

"He's a big game hunter for Mannatorres; we were told never to fuck with him, but we laugh at him because he wears fucked up clothes! This grown-ass man wears clothes from the GAP! Stupid mothafucka!"

"The Gap sells adult clothes, dumb-ass! What, you only seen the Baby Gap?" laughed Lonny, then suddenly Hat picked up the bucket and took the hot ice from the room.

"What is in that room on the porch where Mannatorres' guy stands?" questioned Rick. Cross-Eyed Chris looked at Lonny, measuring his words carefully and said, "I don't know."

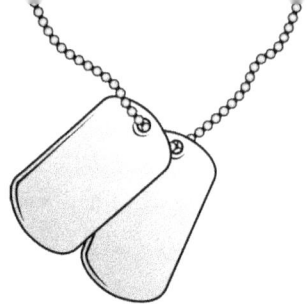

CHAPTER NINE

"You don't know?" Lonny exclaimed, then out of nowhere Lonny took his right hand, slid it inside his pants and started scratching his ass, right in front of Cross-Eyed Chris and everyone else.

"What the hell is wrong with you?" Utah asked him.

"My ass is itching!"

Then with a grin, Utah took his hand and put it inside his pants and scratched his ass and said, "Mine too!" Now everyone was laughing, including Cross-Eyed Chris, so while everyone was laughing Hat came back in with a bottle of urine in the Gatorade bottle and looked at Chris, who nodded yes, he wanted some. Knowing that Cross-Eye's hands were still tied, Hat took the top off the Gatorade bottle and tipped it to Cross-Eyed Chris mouth, so he can drink. The kid took a big gulp to kill his thirst, before spitting out the Gatorade in Lonny's direction, who was laughing harder now. Realizing he just got spit on, Lonny stopped laughing and slapped the shit out of Cross-Eyed Chris, and then everyone was laughing except Cross-Eyed Chris. Utah, who had been laughing and scratching his ass, took his ass scratching hand and placed it on Cross-Eyed Chris face. Lonny followed Utah and placed his ass scratching hand on Cross-Eyed Chris face, with Utah calling Cross-Eyed Chris a butt wipe! Hey butt wipe! Everyone laughed.

Suddenly, JB yelled, "We have company!" The guys who had followed Rick were getting close to the fortress. Lonny placed the gag back in Cross-Eyed Chris mouth while everyone ran to the control room. Utah told JB to cut off the alarm system. The three dummies that had followed Rick were very close to Utah's truck, and all three were looking inside Utah's truck. It became obvious they did not read the big sign in the rear window that clearly says, "DON'T TOUCH THIS!"

Utah's truck alarm system included three heavy duty batteries hooked together. When the alarm was on, the door locks were up, as if the truck was unlocked. The first thief touched the handle and immediately hit the floor. The second thief knelt near the first and touched the body of the truck with his hand, and then he too fell to the ground. The third thief, seeing the first two fall, stopped for a moment, and still

unaware that his boys had received volts through their bodies, decided to go around to the other side of the truck and attempted to get in. He too hit the pavement. Man was that funny! "Are they OK?" Benjamin asked. "They'll sleep there until morning" Utah answered.

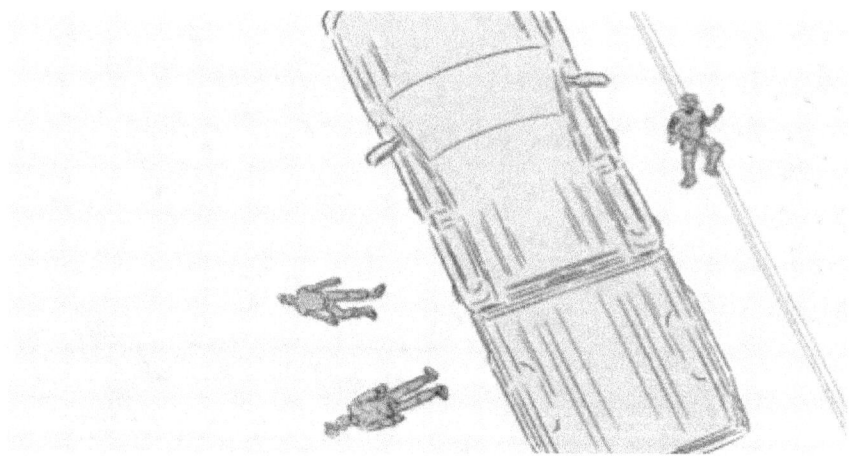

We all made the decision for everyone to spend the night at the fortress; it was too dark and risky to leave for a hotel. We broke out the cots and placed them in the rooms and hallways. Lynnette's cot was adjoined to Hat's naturally. JB put silencers on the mini-machine guns, because of the rats identified by the fortress alarm system during the previous eight hours. With everyone inside, JB placed the system on kill mode to protect us all.

We kept Cross-Eyed Chris in the chair, with the gag in his mouth. We removed his old bindings and replaced them with new bindings on both his feet and hands and placed him downstairs in a room with several rat traps.

Royster had first watch security duty, and less than an hour into his duty a rat trap went off, so Royster went downstairs into the basement where Cross-Eyed Chris was being held. He looked at the rat with its head half off, then looked at Cross-Eyed Chris, his eyes were big as dinner plates. Royster sucked his teeth and told Cross-Eyed Chris "You should have seen the size of the rats in our bunkers in Iraq" then smiled and closed the door. Royster heard Cross-Eyed Chris whimper loudly as he went back upstairs. Outside, the mini guns were putting massive holes in those rats identified by the security system.

As daybreak came, Garries relieved Royster and cut the alarm system off after checking the area. He then checked the four 5k generators that supplied electricity to our fortress. Just like Iraq, the guys urinated in Gatorade bottles - everyone except Lynnette, that is. We used bottled water to wash up and brush our teeth, because Maurisee and JB hadn't figured out how to extract water from the shutoff water line. Hat, Lynnette, Utah, and I decided to get breakfast for everyone, so Hat took three bottles of water with him as we exited the fortress. Utah cut off his truck alarm, while Hat poured the bottles of water on the dumb, dumb and dumber thieves. Once cognizant, they ran away, and Utah did his Chris Berman impersonation, "There they go rumbling, bumbling, stumbling, whoop!" as the thieves staggered away down the street, but in the back of my mind, I was thinking, "They will be back."

Hat and Utah argued about Hat driving to McDonalds, with Hat telling Utah to let him drive and Utah refusing and asking Lynnette to give him directions to McDonalds. As we were going through the drive-thru, Lynnette said hello to one of her friends working the drive-thru window.

"Girl have the bathrooms been cleaned?" Lynnette asked, and the pretty sister working the window confessed they had indeed, so kissing Hat, Lynnette told him she would be right back and slid out of the truck. Hat followed Lynnette with his eyes and scanned the inside of McDonalds, while the fine sister at the window unexpectedly asked Utah his name. smiling and stumbling over his own name, "Ut… Ut… Utah!" he said, like he had a speech impediment. Hat laughing, leaned over Utah and asked the girl "what time do you get off?".

"Four," she said shyly, and then Utah, now feeling his bravado coming back, asked if he could pick her up.

"Yes," she replied, and now I was in shock, because this sister was gorgeous, and Utah was truly ugly and and is from Utah and has no game. Hat nudged Utah to get her number, so Utah said, "Can I get your number?"

"Lynnette knows my number," she replied, and then Lynnette jumped back in the truck with the food and drinks.

"OK girl see ya!" Lynnette waved, and the fine sister in the window told Lynnette to give her number to Utah, as Lynnette seemed surprised. "Karen girl are you sure?" she asked, and started laughing, so Utah touched Karen's hand and said, "I'll see you at four!" and we all started laughing because Utah suddenly got bold.

As we drove off, still laughing, Lynnette asked Hat what happened while she was gone, and Hat told her, "Utah has mad game, and claimed Karen off of waivers!"

"Well don't mess over her, Utah, she also works at the hospital with me cooking part-time," Lynnette warned him, "She has two part-time jobs with one little mouth to feed, plus she's taking online college classes. She's nothing to play with!" The smile was instantly removed from Utah's face.

Hat quickly changed the conversation and asked what we should do with Cross-Eyed Chris. "Rick followed Chris as he dropped off his little Sister at the babysitter's and followed him to the supermarket where they grabbed him. We need to figure out what we going to do with his car as well?"

Royster had driven Cross-Eyed Chris BMW to the fortress once they snatched him. After passing out the food, we decided to let Cross-Eyed Chris go after we filled his head with misinformation.

Joey and Benjamin carried Cross-Eyed Chris back upstairs and Joey removed the gag from Cross–Eyed Chris mouth and he went off like Fireworks! "You placed me in a room full of rats! How… what the fuck? Those rats were big! What the fuck? You guys are crazy! Let me go! Let me go! Let me go or kill me!"

Lonny untied Cross-Eyed Chris hands from the chair arm and threw his 9mm in his lap, saying, "Kill yourself."

Cross-Eyed Chris looked at the gun for a moment, then whined, "If I pick this gun up, you're going to kill me!" Lonny taking his gun back from Cross-Eyed Chris said softly to Cross-Eyed Chris ear "I wanted to kill you the moment I saw you." "Listen, shouted Hat, we don't have time for this! We need money; we need to rob and kill Sal today; we don't have time for this! Hold him down!" So, Benjamin and Joey held him down, while Hat placed the 9mm in Cross-Eyed Chris right hand, then removed it and placed the gun in a plastic bag. Lonny untied his feet from the chair and told Cross-Eyed Chris to get up and leave. Still suspicious, Chris didn't move and asked, "What are you going to do with that gun?"

"We're going to kill Sal with it, now it has your fingerprints on it," Hat told him.

As Chris jumped up, JB, who had now ghosted Chris' cell phone threw the cell phone at him and told him, "Here, call the Police."

Chris caught his cell phone and placed it in his pocket and was promptly given an incentive to leave with the placement of Lonny's foot against his ass. Cross-Eyed Chris looked at his BMW as Lonny said "Wait, there is No Way, I am going to let this butt wipe go without whippin' his ASS." Maurisee said "brah, we have done enough to this kid. His fingerprints are on the gun, we got him." Naw, he needs his ass whipped for all those times an elderly black woman told him to pull-up his pants and he cursed her. He needs his ass whipped for all those times an elderly black man said son, pull your pants up and he told them to fuck off." Then, Lonny takes off his belt and begins to whip Cross-Eyed Chris. After about one minute, Maurisee said "Let me get some." Lonny replied "Naw brah your white, this is no time for a time out. Look at em, all this talking is giving him a break." "Do you really think the black youth are the only ones walking around showing their ass?" Maurisee said. Then he took of his belt and accompanied by Lonny, gave Cross-Eyed Chris a good ass whipping. Marty broke up the fun and   set a very bruised Cross-Eyed Chris on his way. Crying, Cross-Eyed Chris looking at his car said "what about my car" "We're keeping it,

thank you. Dumb bastard!" Lonny replied. Marty and Rick came up with a plan of action for the day, and AB, who was still with us but not participating said, "Hey, let's go to Ground Zero!"

"Yeah, I'd like to see Ground Zero," Utah said but Hat replied, "Man, we don't have time for this!"

Marty asked Hat to come outside. "Listen, Hat, we're going to find out who killed One and when we do, he fades to black, he'd be a memory. In the meantime, let's take everyone to Ground Zero, and we can go over and finalize the game plan while in the vehicles, then meet back at the fortress with all the information collected, and then execute."

Hat agreed, so we all headed out towards Manhattan, but prior to departing, Maurisee reached into his bag of tricks and pulled out earpieces and handed them to everyone. "Are these earpieces for our cell phones?" Royster asked. "They look like cell phone earpieces, but they're limited range, one touch, voice-activated receivers and senders," Maurisee replied. "What?" Marty frowned, then JB pushed the only button on the earpiece and said, "Marty, hey you!" Marty heard JB and replied, "Hey you!" But JB didn't hear anything, so he warned Marty, "You need to push the button, say my name, then talk. Every time you talk, push the button and say the person's name and remember, you don't have to yell because the earpiece picks up a whisper." After everyone finished playing with the new gadgets, we loaded up in the Suburban's. I was riding with Lonny, Utah and Lynnette, with Hat driving. The second Suburban had JB, Maurisee, Garries and Barnett,

with Royster driving, and the third Suburban had Benjamin, Cousins, Joey, and AB, with Marty Driving. Rick took the keys to Cross-Eyed Chris' seven series BMW and drove that car to a chop shop on Linden Blvd. Money in hand, Rick jumps in the Suburban with Marty. Once we arrived in Manhattan, we made a detour and stopped at St. Jude's Children's Research Hospital. Rick jumps out and runs into the emergency room entrance. Minutes later, he returned with a woman, the Head of the Children Cancer Research Department. When Rick told her he was donating the money to the hospital, she hugged Rick and started crying, but Rick just jumped right back into the Suburban with Marty and off we went to Ground Zero. A sense of accomplishment fell over the group for some reason. The intense faces disappeared. Hat feeling good, turned up the old hip hop music we were listening to. Utah quickly attempted to change the radio station to Country, which caused a lighthearted conflict. My own thought process was a simple one; if this is going to bring the guys together, we should go and steal all the drug dealers' cars and give them to Breast Cancer Research, Soups Kitchens, Aids Research, the Jimmy V Foundation, the Red Cross, the Wounded Warriors Foundation, and the Old Soldiers Home. Lonny, who had become very suspicious of Rick, muttered, "Who does he think he is - Robin Hood?" But nobody answered him as the radio station fighting continued, with Lynnette chipping in. Maybe no one else heard him, but I did. Everybody was all smiles as we exited the vehicles, but those smiles took a lunch break as we got close to the gigantic hole in the ground that used to be the World Trade Center. The pictures along the fence we were leaning against told a story like no other in my lifetime. Hat, who was holding the hand of Lynnette, seemed to be fixed on a picture of a little girl with her Mother crying. Utah had his eyes on all the construction going on. Everyone seemed to be concentrating and contemplating something. Finally, Utah, who was standing on the other side of Hat, broke the silence by calling to Marty, Rick, Benjamin, AB, and Joey, then motioning them with his hands

to come to where he was standing. When they arrived, Utah asked, "What does that construction sound like to you?" Although he didn't ask me, I thought the sound sounded like a muffled blast. When Utah said that the sound reminded him of the blast that took his legs, the heads of the guys all dropped. Time was frozen, and the church nearby chimed a familiar tune, letting everyone know it was high noon. Hat hugged AB because of the tension between them, but also because it was AB that helped Hat on that terrible day when Utah got his legs blown off. It happened on their second tour together in Iraq. Three motorcycles carrying a passenger with AK47's was circling around the convoy, when without telling anyone, Utah shot two of the motorcycles, causing them to crash into the oncoming traffic. Utah, Hat, AB and Joey (who was driving the lead vehicle in the convoy) celebrated the shooting, and Utah even told Joey to stop the HUMVEE so he could look at the mess he just made. Joey stopped the vehicle and Utah jumped out and raised his M4 over his head, yelling and screaming, "Yeah, yeah, yeah!"

Hat told Utah to get back in the Hummer, and AB yelled "Come on let's go! Damn he's hard-headed!" but Utah was still bouncing away from the vehicle and was 15 feet in front of the HUMVEE and 5 feet inland when he stepped on a land mine. Blood flew everywhere as the horror engulfed Hat, AB, and Joey. They all jumped out of the HUMVEE, as Hat reached Utah first. Hat quickly took his ready-to-go tourniquet out and strapped it onto Utah's upper right thigh.

AB followed suit with a second tourniquet to Utah's upper left thigh. With his eyes closed tight, Utah began to sing, as Joey called for a medic. Joey jumped out of the HUMVEE and ran to Utah, picking up one of Utah's mauled legs and attempting to pour water on it. Hat waved him off, so Joey stood up yelling, screaming and waving for Marty to drive faster, Marty was in the trail HUMVEE and was stuck behind the traffic jam Utah caused minutes earlier. Joey then posted himself as security but continued to look at Utah. Hat took Utah in his lap and sang the Toby Keith song with Utah,

"Well a man come on the six o'clock news says somebody been shot, somebody been abused, somebody blew up a building, somebody stole a car, somebody got away, somebody didn't get too far yeah, they didn't get too far"

AB switched places with Hat and continues singing, so Hat could give the ambulance UH1H directions to land. Not waiting for the UH1H to land, Hat reached into the chopper and retrieved a litter, on which he placed Utah and his legs which went into cold packs. Hat and

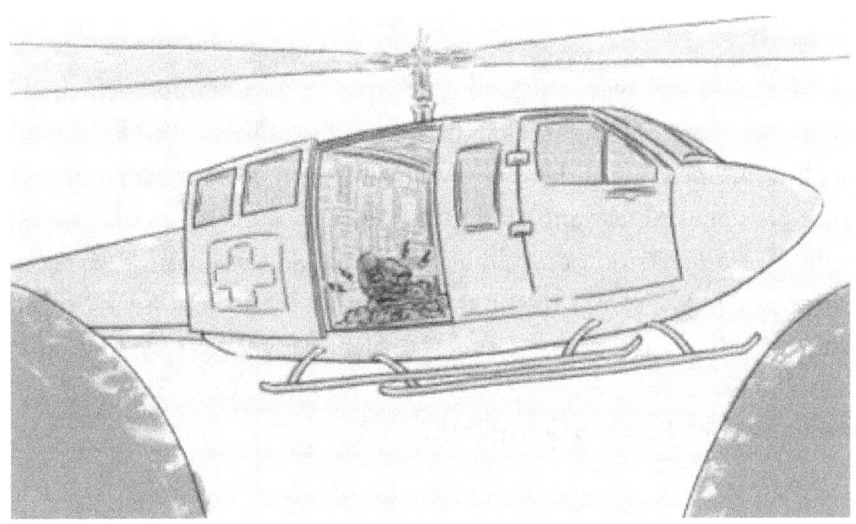

AB carried Utah to the chopper, and then Hat and AB motioned the chopper to hurry and get off the ground. Utah was being strapped in as the chopper took off. As the chopper swung around, everyone could plainly see the onboard medic using a defibrillator on Utah.

After thinking back to what happened in Iraq, Utah took a deep breath and said to the guys, "I've never thanked y'all for saving my life; therefore, there's no need to thank you now!" Lynnette was crying as we all hugged displaying man love. After the group hug, Joey, Benjamin and Barnett got into the Suburban and headed to Brooklyn, the Canarsie section, to meet and greet Mannatorres.

JB, Maursiee, AB, Garries, Cousins, Marty, Royster, and Rick got into the second Suburban and headed back to the fortress. Lonny, Hat, Utah, Lynnette, and I left the 9/11 site and made for the Police Station to view the crime scene pictures of One's death.

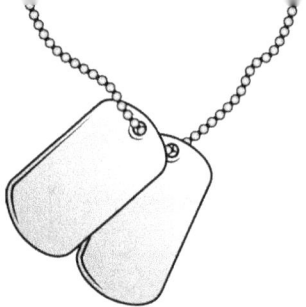

CHAPTER TEN

Joey Pisciotta was the coolest Italian alive, and he knew we had to deal with Mannatorres, and deal with him on a level Mannatorres understood. Joey drove by the private school that Mannatorres' seven-year-old daughter attended and pointed out to Benjamin and Barnett the car that contained Mannatorres pickup men. The two men had only one mission; to pick up Mannatorres' daughter and walk her to the ice cream shop before taking her to the restaurant where Mannatorres was waiting at, his restaurant.

Joey then drove a block away from Mannatorres 'restaurant and got out of the truck, leaving instructions for Benjamin to follow to the letter. Benjamin tap-coded Joey 20 minutes later, and Joey walked one block to Mannatorres' restaurant, walked in and asked to speak to Mr. Mannatorres. He was searched by two big brutes who took his mini monitor and cell phone, before Mannatorres motioned for Joey to come forward. Joey stopped four feet away, and Mannatorres said, "I know your family," (in Italian). Joey responded back that he doesn't speak Italian – (in Italian).

In English, Mannatorres asked, "So what can I do for you, kid?"

"Sir," Joey replied, "I'm sorry to inform you but my friends and I have been working against you in the Brownsville section, and we apologize for the inconvenience. With your permission, we would like to kill some of the people that work for you. We don't want to kill the Italian spy you have stationed on the porch, but we would like to kill anybody and everybody that stands in our way. We're looking for the person who killed somebody of no importance to you."

Mannatorres laughed and waved to the two brutes, "Get him out of here!"

With a sense of urgency in his voice Joey said, "I need you to listen to me! Cut on my monitor." One of the brutes retrieved the monitor and handed it to Mannatorres, who cut the monitor on to see an unsteady camera picture of his daughter eating ice cream. JB had mounted a mini camera on Benjamin's belt, about the size of a dime.

"I need my cell phone," Joey said, and a stunned Mannatorres motioned to the other brute to release Joey. Joey then tap-coded Benjamin to pan over, so Mannatorres could see Barnett standing behind one of the pick-up men. Barnett seemed to have a gun on the unhappy man.

"Pan back on my daughter?" a distressed Mannatorres asked.

"Sir, your daughter will not be harmed," Joey replied, "this is her normal pattern; everything is normal. But I need you to listen, Sir."

Mannatorres continued to watch his daughter eating ice cream, "We would like your permission…" Joey began but Mannatorres interrupted him from finishing his plea.

"You hold my daughter captive; you threaten my family, and you want my permission? This is a family matter; why are you involved?"

"Because this is a family matter," Joey replied.

"Let my little girl go," Mannatorres said, but Joey just told him, "When she finishes her ice cream." Then Joey tap-coded Benjamin to give the little girl another scoop of ice cream. The camera moved as

Benjamin ordered another scoop and placed the scoop in front of the little girl, who said, "Thank you!" and began eating her vanilla scoop out of a bowl.

Mannatorres, now confident his daughter was safe, asked, "When?"

"We'll be finished by tomorrow - before 4pm" replied Joey.

"And the product?" queried Mannatorres.

"I am afraid some of it might be wasted," Joey warned.

"So, you want me to lose money? For these moolees? Come on, they eat their own kind! They feast on one another!"

"Yes, and you feed them!" Joey told him.

"That's because they're hungry, they want the fast money for the fast cars and the Jordan sneakers. What are you kidding me? They go from school to prison, NO! do not touch my money."

"You are full of SHIT" said Joey."

"Soldier boy, you want to start with me? The great Mannatorres? I worked my way up from the sorrowful docks to build my empire. I killed a lot of people on my way up - I could have you and your family crushed just by winking; you piece of shit. You're a memory; you're nothing to me. I am Mannatorres!"

Joey interrupted him, "Sir, your daughter is almost finished, so should I have her eat another scoop of ice cream, maybe chocolate this time? Or should we give her strawberry?" As he began to tap-code Benjamin.

Mannatorres turned off his motor mouth and cried, "No; no more ice cream! She might get sick," so Joey tap-coded Benjamin to have him and Barnett walk away.

Seeing this, Mannatorres asked, "And my daughter?"

"My guys will follow your guys," replied Joey.

Mannatorres paused at this, and said, "The man responsible for your troubles will be present at noon tomorrow. I can always find another Moolee to replace a dead Moolee."

"I thank you Mr. Mannatorres," said Joey quietly, "you are as wise as people say. We ask forgiveness." He then tap-coded Benjamin with information on where to pick him up.

A few minutes later, Mannatorres' men pulled up at the restaurant, and Mannatorres quickly opened the door to the limo and hugged his daughter with one hand, while slapping the two escorts with the other hand.

Joey meanwhile, walking backwards, was picked up by Benjamin. "That was too easy", Benjamin crowed.

"Man, that was just a cake walk!" retorted Barnett. "The stupid, big, dumb-asses didn't even give us a second look when we walked up on them; they were too confident in their own environment."

"Yeah, just like us, right? Now, do you think we got away with something?" Joey asked. "Mannatorres is a DON! Do you think he's going to let us get away with that? Do you think he's not going to Fire back? Check behind us to see if we are being followed!"

At this, Benjamin who was driving made a left turn and Joey, Barnett and Benjamin looked behind for vehicles following them.

Once they arrived at the fortress, Joey professed the mission was complete. "I followed your instructions down to the very last detail, Rick!"

Rick smiled and said, "Well that's a first for you, Joey!" as Maurisee, JB and AB laughed.

"What?" demanded Joey while he, JB, Barnett, Maurisee, Marty, Rick and Royster loaded up the Suburban, "I always follow instructions!"

"What about the reporter?" Maurisee asked him.

"The reporter?" asked Joey.

"Yeah, the reporter," Marty remembered.

It was our second tour in Iraq, and Hat was told by Lieutenant Ward that a reporter was going to be embedded on a mission with us. Hat came back to the HUMVEEs and told everyone that a reporter from some network would be embedded with us for our mission, and we were not to speak with this individual. "Don't say hello; don't say shit. Reporters will never tell your story; they will only tell their story - these people have a hidden agenda, and despite your intentions, they always get the last word. You get the last word if you never speak."

AB asked him, "What if it's Geraldo Rivera?"

Hat said "Hell no! Don't say shit!"

"What about Christiane Amanpour?" Benjamin asked.

Hat and Marty both warned him, "Hell no! Don't say shit!" especially to her!

"NO, NO, NO! Don't say shit, I got it, Said Joey."

"I say again," Hat said, "DO NOT TALK TO REPORTERS!"

"What about Jason Carroll?" Benjamin asked. "Mr. Carroll is the real deal, honest, and trustworthy, same with Shepard Smith."

Yes! He's good people and so is Kyra Phillips, she's a real patriot, a true American that understands the extent of our sacrifice, I think they both do, but that's it!" Marty said.

After a moment's thought, Marty asked, "Wait - what about Sports Center reporters?"

Hat pauses and said, "Yes, OK, if the individual is from SportsCenter talk to them, but it must be Chris Berman, Stuart Scott, Trey Wingo, Kenny Mayne, Tom Jackson, Chris Mortensen... and who else, Marty?"

"Bob Ley, Chris Fowler, Stephen A. Smith, Scott Van Pelt, Linda Cohn, and Bonnie Bernstein. OK?" Marty replied. "Ok," Hat said, "So if the reporter is from SportsCenter talk to them; other than that, don't say shit. DON'T SAY SHIT. DON'T SAY SHIT!" Joey pulled out a piece of paper as he walked away, smiling. About an hour later, a country tobacco-chewing cameraman walked up to Hat and announced, "Hi! My names' Troy: I am the cameraman. I used to be in the Army. I spent four years jumping out of planes at Fort Bragg - HOOAH! I'm going to tell your story to the folks at home, using this thing here" pointing at his camera, "So what's your name?" Hat gave him one of

those Tilden Projects Fuck You looks, and then walked away, not saying shit. "OK," said the cameraman. Then the reporter came out of the TOC where he had been getting a briefing from Lieutenant Ward. They decided to set up right in front of Marty's HUMVEE. He began his reporting, then moved over and asked Marty where he was from. Marty looked right into the camera and said nothing, so again the reporter asked him, "Where are you from?" Again, Marty said nothing. Laughing outwardly and boldly looking into the camera, Marty said " you are a motherfucking goat sniffing dog" and just like Hat did, Marty

walked off. "Cut!" yelled the reporter then went back into the TOC, where he told Lieutenant Ward, "Your men don't seem to want to talk with us." We're attempting to tell a story here that millions of viewers will be interested in; we need some cooperation, Captain." Lieutenant Ward said, "I'm a Lieutenant," in a sarcastic way, then poked his head outside the TOC and called for Staff Sergeant Hadley. Hat walked into the TOC and stood stiffly at attention. The Lieutenant told the reporter to step outside. Once they were alone, the Lieutenant told Hat he WOULD talk to the reporter.

"Sir," Hat said, "I ain't sayin' shit to that blood-sucking, bullshit, double-talking low life reporting bastard. I just don't know who to trust; they can edit the interviews and make you sound like an idiot with those sampling sound bites. My job does not entail babysitting some candyass, self-centered liar who talks bullshit! I'm not saying shit to that asshole reporter. Reduce me in rank; send me the fuck home, because I am not saying shit. We have Public Affairs Officers designated to kiss his ass; I'm not doing it! Now, I need to get back and

prepare for this mission, SIR!" Lieutenant Ward told Hat to "at ease." "Are you disobeying a direct order?" "Have I ever disobeyed a direct order?" Hat asked. "So, think about this Sir -Do you really want me to represent this unit by opening my nasty filthy gritty New York mouth and speak to the millions of wonderful wonder-bread eating, 4th of July flag waving Americans? Or would it be appropriate for the Public Affairs Officer to speak with him? With the LT still looking perplexed like he still didn't understand, Hat said, "OK Lieutenant bring that sorry lowlife Sonbitch reporter in here and I'll say some REAL pleasant things to good ole' Ma and Pa America."

Pissed, the Lieutenant told Hat too, "Get the fuck out of here!" Leaving the TOC, and telling the reporter and cameraman to follow him, the Lieutenant walked right up to Joey and said to the reporter, "He'll talk. He always talking, right?" Joey shook his head, yes, so the reporter and cameraman set up, and then the reporter began with Joey standing right next to him.

"We're in a secret location in Iraq with a bunch of Soldiers enthralled in a battle to survive. We have with us… state your name, Soldier?"

Joey didn't say a word, so the reporter said, "These guys are shell-shocked! So, what's your name Soldier?"

Joey looked at Hat, Marty, Maurisee, JB, and the rest of the squad which had gathered for his Television debut. Joey then removed a piece of paper from his pocket then looked at the reporter and asked, "Are you Chris Berman or Stuart Scott?"

Confused, the reporter said, "No."

Pulling out another piece of paper from his pocket and reading, Joey asked, "Are you Trey Wingo, Bob Ley, Scott Van Pelt, Kenny Mayne, Tom Jackson, Linda Cohn or Bonnie Bernstein?"

"Wait", the reporter said, "those individuals are from ESPN!"

"Yes, exactly," Joey smiled, "are you from ESPN?"

"No" the reporter replied.

"Well then," said Joey, "FUCK YOU, I AINT SAYING SHIT!" Then he walked away.

Everyone cracked up, but the Lieutenant was so mad he called Joey Pisciotta over and cursed him out, and then he told everyone to go back to preparing for the mission. The Lieutenant gave the interview to the reporter, with only the cameraman as his audience.

"Man," Marty told Joey, "Man, the Lieutenant stayed mad at you. Hat told you not to say a word; man, was that shit funny?"

That tight-ass reporter never came close to us again.

Maurisee went over the next phase of the plan and Marty decided to keep Cousins at the fortress to ensure incoming intelligence was recorded. Rick and Maurisee had built a blueprint for the heisting of two cars driven by the drug dealers. Benjamin and Barnett were to steal the vehicles and drive to where JB would use a truck to momentarily block the pursuers, so they detoured down a path where they placed spikes in the road. This would cause the motorcycle chasers and car chasers alike to receive flat tires.

Royster, now in a jumpsuit the color of the beam he would be leaning against, would pay to get on the train platform and lower himself on the beam with the remote fly control. Royster's mission was to wait until the door of the drug location was open and fly the remote-controlled camera fly into the room Rick believed had the drugs. Maurisee would park close to Royster's location once the chase started and using the unmanned aerial system, Maurisee would signal to Royster when the door was open to fly the remote fly into the room. Rick would blow up a car, diverting the scooter boys' attention, so that Benjamin and Barnett could steal the cars.

Everyone acknowledged their roles, and Benjamin and Barnett started arguing over who should drive the Mercedes and who should drive the BMW. Marty dropped Royster off at the train station and JB at the truck rental location. Rick placed the explosive package under a car, while Maurisee, Benjamin, and Barnett waited down the street for the car to explode. Maurisee tap-coded JB, Royster and Rick to ensure everyone was in place.

Once everyone was in place, Maurisee employed the unmanned aerial system, Marty gave the signal and Rick remote-detonated the car bomb. The motorcycle kids immediately turned their attention towards the explosion and walked slowly in that direction. Royster lowered himself down 15 feet, to a position on a beam supporting the train platform. Maurisee slowly drove down the street and dropped off Benjamin and Barnett at the drug dealers' cars.

One young kid saw Benjamin and hit the button on his cell phone, and one of the guys in the room came out; he was the alert man. Royster flew the remote camera fly into the room, and then from that moment on Maurisee was recording everything from the fly-on-the-wall. Benjamin and Barnett drove off, and the motorcycle kids and two other drug dealers gave chase in cars.

Maurisee told Royster to fly around to get all the rooms inside the place, but one of the bikini-wearing females told another girl a fly was in the room, so Maurisee told Royster to guide the fly under the black leather couch, to camouflage the unit.

Meanwhile, the car chase was in full gear as both the motorcycle boys and the two cars gave chase after Benjamin and Barnett. Marty followed the chase to ensure nothing went wrong during this operation.

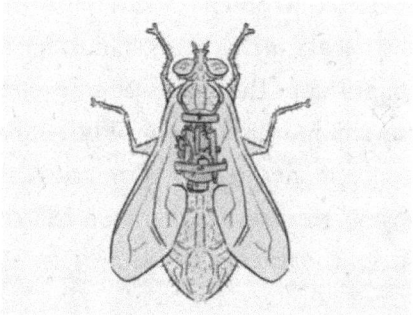

Benjamin, who is in front of Barnett, told JB that they would be there in one minute. Benjamin and Barnett then passed JB, who pulled up in the rental truck, causing the pursuing motorcycle riders and car chasers to detour. Moments later, they hit the spikes in the road and the motorcycle riders flipped off their bikes, and the car's tires were blown.

Back on the block, Rick joined Maurisee in the Suburban. Maurisee had seen enough, and he told Royster to get the remote fly-on-the-wall out of there. One of the guys in the room received a phone call from one of the lieutenants who failed in the chase. He opened the door to get a better reception on his cell phone, giving Royster more than enough time to fly the unit straight out and back to his location. They observed another drug dealer get into his car and drive towards the location of the failed car chase.

Royster lowered himself to the street and was picked up by Maurisee. Benjamin and Barnett drove their cars to the chop shop on linden Blvd. Maurisee with Royster in tow, picked them up. Maurisee drove to a soup kitchen. Benjamin got out of the vehicle and went inside the soup kitchen. Moments later, Benjamin exited the soup kitchen with a woman chasing him with a handful of hundred-dollar bills. Benjamin smiled as he was being hugged tightly. The next stop on their way back to the Fortress was a Battered Women's Shelter. Barnett walked into the Battered Women's Shelter and the scene repeated itself except a man was shaking Barnett's hand repeatedly as Barnett walked towards the Suburban. The man was so excited, Barnett had to help him pick-up several hundred-dollar bills he dropped while shaking Barnett's hand.

The men returned the truck and headed back to the fortress where Maurisee, Garries, Royster, Marty, and Rick analyzed the information from the fly on the wall.

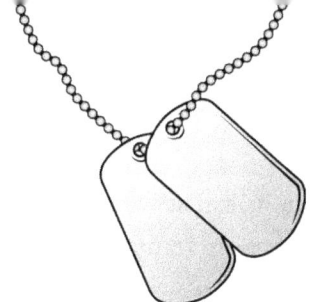

CHAPTER ELEVEN

As we were walking to the Suburban from the 911 site, Lynnette opened the back door for Utah and told him to jump in, "I'm controlling the radio now – I've had enough of Toby Keith, Utah, don't you like anyone else?" she asked. Lonny chuckled and laughed while helping Utah flip over the chair. Why Utah decided to flip over the seats instead of flipping the seat down, I will never know. Lynnette, now in the front seat, found a nice old head radio station. Chaka Khan was singing.

"I will love you anyway, even if you cannot stay,
I think you are the one for me, here is where you ought to be."

Lynnette was singing to Hat, trying to take his mind off the task at hand, but Hat seemed distant, different, kind of like in light fog, he seemed off but maybe he was just thinking about his cousin, One? Still singing, she hugged Hat, leaning to whisper into his ear at which he smiled and nodded.

Lynnette made a phone call as we drove towards the Police Station, and Hat announced we needed to make a detour. As we pulled into the parking lot of McDonalds, I saw that beautiful woman working the window again. Lonny, who wasn't present when we went to McDonalds earlier, voiced his approval of Karen by saying, "Oh, my damn, she's fine!" I turned and looked at Utah, who had a shit-eating grin on his face. Karen greeted Lynnette at the window of the truck, and at the same time Lonny hurried to get out, opening the door for Karen.

Once Karen was in the Suburban, she hopped over the seat and sat next to Utah in the last row. Lynnette introduced Karen to Lonny, as Lonny seemed stunned that Karen chose to sit next to Utah. Karen and Utah whispered, laughing loudly. Cheryl Lynn's "Got To Be Real," was playing as a dejected Lonny said, "No shit, it's got to be real." Hat, Lynnette, and I laughed as the whispers continued in the rear of the Suburban.

The mood in the truck changed as we parked near the Police Station. Walking in formation fashion, Hat led with Lynnette holding his hand, Utah and Karen behind them, and Lonny and I brought up the rear. Once inside the Police Station, Hat asked to speak with Officer Nelson.

I got a cold feeling sitting on that bench 15 feet from the elevated reception desk, and I felt the whole place seemed cold, so I decided to stand up. Officer Nelson arrived with the pictures but decided to motion Hat outside, so we all followed Hat outside.

"There are three youths in the Police Station right now, blaming you for the murder of Chris Walker!" Nelson said.

"I don't know a Chris Walker, and I haven't killed anyone; I've been with her" Hat said, pointing to Lynnette.

"They call him Cross-Eyed Chris," Nelson added.

"OH SHIT!" blurted out Lonny as Nelson continued.

"These three youths said they saw Cross-Eyed with you, and when they tried to rescue Chris from your grips, you hit them with a stun gun. A warrant for your arrest will take place in 24 hours."

Then Officer Nelson pulled Hat away and walked a few steps and told Hat, "Mannatorres has a lot of friends in this precinct. He wants your boy Joey dead."

Hat looked at Nelson and said, "A lot of friends?"

"You have 24 hours!" was all Nelson said, giving Hat five pictures of the crime scene.

Hat rejoined the on-looking group that had formed a tight circle. As the pictures were being passed around, Karen said she knew the girl in the picture. "She worked for Leon," she stated. Karen told us that Leon asked girls to work for him, making crack on the first floor of the projects. The girls work in bikinis and Leon pays them well.

"This girl is from our block," Karen said, "she worked for Leon but later became his girlfriend. I heard he fronted her money to go to college and even put her up at a Manhattan apartment."

Hat wasn't listening to Karen, but he kept looking at the picture of his bloody cousin whose face was a mess. Hat looked at me and said, "TAP TAP TAP – TAP…..TAP TAP TAP – TAP."

I looked at Hat and said, "L."

Hat showed me three pictures of his Cousin One, and in each he was displaying three fingers with two fingers folded on one hand and one finger displayed with four fingers folded on the other. "Now I'm almost sure Leon killed One," Hat said. "Let's go back to the fortress."

Hat had forgotten about Lynnette and was walking fast towards the truck, intent on picking up his gun and going to work on Leon. He drove like crazy trying to make it back to the fortress, as Utah told him to slow it down. Lynnette was looking worried and without saying a word, she touched Hat on the shoulder. Hat, who took up a position by leaning forward to drive fast, leaned back into the driver's seat after Lynnette touched him and returned to driving normal for a New Yorker.

Stopping at a red light, Lynnette looked at Hat and turned down the radio (which had a Rob Bass song playing) and said quietly to Hat, "Marry me, James Dayshawn Hadley; marry me now."

Hat looked at Lynnette all stunned, and asked her, "What did you say?"

"I want to get married," Lynnette replied, "I want to be your wife, and you, my husband."

Hat went from a frown to a smile but did not say a word.

Minutes later, we were at the courthouse, which was five minutes from the Police Station and ten minutes from closing for the evening. People were exiting the courthouse as it was nearing 1800 hours, but we made it to the marriage Judge in the nick of time. The Judge conducted the ceremony, and when he finally asked, "The ring?" no one had a ring, but a giggling Karen stepped forward and gave up one of her rings; the woman had a ring on every finger. It took Lynnette three tries before she found one of Karen's rings that fitted her ring finger, then Hat put on the ring, the Judge pronounced them man and wife, and Hat kissed Lynnette passionately.

Lonny and I were still amazed at the turn of events as we walked back to the truck, and Utah and Karen were giggling as usual as they walked. Hat, who was all frowns about ten minutes ago, was extremely happy as he hugged Lynnette while walking towards the truck. Lynnette kissed Hat on the cheek and Hat stopped in mid stride and said quietly, "Thank you and I will never leave you. I love you and Brice, and we're official now."

Hat's smile continued as he drove calmly down the street. His face once again turned cold as we arrived at the fortress, where he left his wife behind as we exited the truck, making a beeline to the control room. Karen entering the fortress looked around like she'd never seen anything like the fortress. In the control room Maurisee, JB, Garries, Joey, Marty, Rick, Benjamin, Barnett, Royster and AB were all gathered. Hat walked up to Marty and Rick and showed them the pictures.

"This confirms it," Marty said, "3/1 is 'L' for Leon. One left us a clue before he died. Leon wants us so dead he paid this dumb-ass kid five thousand dollars to kill any one of us. We have the dumb-ass tied up downstairs."

Hat looked puzzled, so Marty explained, "the stupid kid just tried to walk up on us, and he shot at the fortress Wild West style!"

Royster butt stroked him, "I just exited out of the window and flanked him. Tough little guy!" Royster said.

"I shot him with truth serum (Sodium Pentothal) JB gave me and he started singing like New Edition."

"Fuck that," said Hat, "Leon is going to get a hot one."

"About damn time!" Lonny exclaimed.

"All of this is bullshit, secret squirrel bullshit. I'm going after Leon!" said Hat, looking at AB. He turned to Joey, "What time will he be at the projects?"

"Mannatorres said 12 noon," Joey told him.

"Well, that's where I'll be tomorrow."

"What about Mannatorres?" Asked Rick

"Man, what is with you?" Lonny asked, as he stepped towards Rick, invading that imaginary safety circumference around any individual. "You've been pulling our strings ever since we arrived and even before then - why?" Marty went to step between them, but AB pulled Marty back, and Lonny continued.

"Who stayed in contact with Marty while we were in Iraq you? Who made the arrangements for Hat's rental car you? Who called Hat and told him not to go to the funeral, you? Who had the pictures of the drug dealers, you?

Royster, Benjamin, Cousins, and Barnett moved in closer. "Who has been steering this whole thing, you? Who wants Mannatorres dead, you? You FUCK!" and suddenly Lonny punched Rick in the face.

Rick hit the floor but didn't get up to fight back, taking his time to stand up before responding simply, "You have a real good punch," as he wiped the blood from his mouth. Looking at Marty and Hat, Rick said, "I came home after our second tour in Iraq because my father had a stroke, and because after Utah lost his legs, I became frightened. I couldn't sleep - I had to get out of the Army, I needed to leave Iraq."

"So, I immediately had to take over my father's dry-cleaning business. Two weeks later, two of Mannatorres' men walked into the dry cleaners, demanding I turn over the monthly protection payment. One of Mannatorres' men told me my old man kept an envelope in the safe, marked with a red line. I was confused, so Mannatorres' guys told me to open the safe; inside were six white envelopes, each with a single red line on them." "Mannatorres' men reached over my shoulder and took one of the envelopes and left. I watched them go to the candy store, I also saw Leon was waiting at the corner for them and I saw him give them money also. So, I closed the dry cleaners and went home to confront my father. My stricken Father could barely speak, and he seemed upset over my questions. He seemed angry and struggled to speak. I told him I needed answers; what was going on and how long had he been paying Mannatorres?"

"My Father told me that years ago, Mannatorres came to the store late one night and told him he would be a very rich man, and then walked out. Mannatorres then burned down every dry-cleaning business within 15 miles of our store. The business increased by 400 percent. My Father also told me about the night your Aunt Lacy died.

He told me that Mannatorres was in the store late one night collecting money, when they heard a car crash. My Father and Mannatorres ran out of the store towards the crash, and when they arrived, they found your Aunt halfway through the front windshield. My Father ran back to the store and called your mother, and then the police, and then quickly ran back to the crash site. Moses Jackson, the owner of the bar, was also involved in the accident, he was in the other car, a head-on collision killed your Aunt."

"Moses stumbled to get out of his car and began crying when he saw your Aunt. He was asking my Father to help him find his son, your Cousin One was on the floor in the back, unharmed. Once Moses saw his son was safe, he tried to pull your Aunt through the window, but she died on the screen. Your Mother arrived and screamed as she tried to rescue her sister. Crying and sobbing she snatches One from Moses and hugs him tightly. My Father thinks the rain caused the accident."

"Moses Jackson is Leon's Father, and One's Father. They are your Kin Folk. Ten years later, Mannatorres burned down the bar Moses owned with Moses in the bar and fronted Leon for the drug empire when crack was running wild. So – you'll be killing your cousin, and I'll be killing Mannatorres." Everyone was shocked. Stunned, Royster says "wait, I don't understand." Rick explains while Hat in a state of shock, slowly walks away to call his mother. Moses had a wife and they had Leon. Moses cheated on his wife with Aunt Lacy and they had One. Moses brother was Dirty Burt. Dirty Burt was married to Lynnette's mother. Dirty Burt and Hat's father were in Vietnam together, Dirty Burt came home, but Hat's father didn't." "Wait Wait" Royster said, Lonny interrupts "Leon and One are half-brothers" "They are brothers" Marty shouted. Then Joey added "Lynnette and Leon are blood cousins and Hat and One are blood cousins." Royster says, "So Hat just married his cousin, but not by blood?" "Yes" Rick added.

"So Manatorres killed Moses by burning him in the fire and fronted Moses son Leon to run the drug situation. Leon kills his uncle Dirty Burt and his brother One, now Hat wants to kill his wife's cousin." I'm confused" Royster said. Barnett replies "Do I have to raw you a map! They are all "Kin Folk."

While they all allow these words to rent space in their heads, Hat off in the distance confronts his mother. "Ma," Hat said, "tell me everything about the night Aunt Lacy died." Hat's Mother told Hat everything, and confirmed Leon and Hat were in fact cousins. She also told Hat that Leon might have killed his uncle, Dirty Burt, and that Dirty Burt and Moses were brothers. She warned Hat not to mess with Leon.

"But why would Leon kill One?" Hat said, thinking out loud to his Mother.

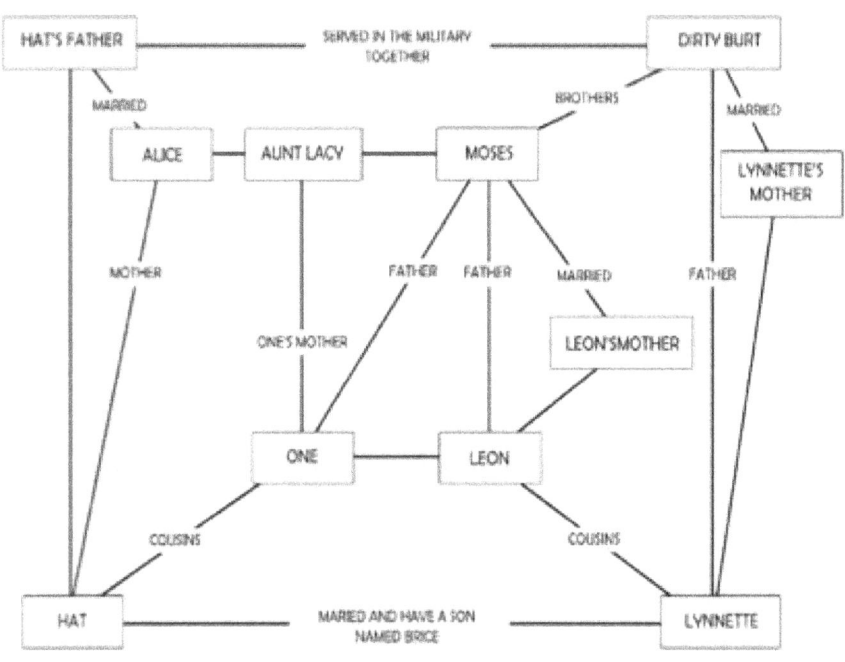

"Baby, I don't know!" she replied. "But I know Leon is crazy."

Hat hung up and Maurisee took the pictures from Hat to have JB scan the pictures into the computer. Then JB blew up a framed picture on the desk in one of the pictures. Looking over JB's shoulder, Karen said the girl was Charmaine who worked for Leon, and that was Leon in the picture. JB said softly, "Hat, it looks like she was dating One and Leon."

Hat asked Royster to drive Lynnette and Karen to the hotel. Taking Lynnette outside, he told her, "I'll be there in one hour - I promise, Mrs. Hadley."

"Hat, it's our honeymoon" she warned him.

"I'll be there."

Karen kissed Utah then stepped into the vehicle. Hat waved as they drove off then turned and headed downstairs where the captured young punk sat, tied up and knocked out. Everyone followed, as Hat asked about the truth serum given to the young man.

"It's experimental, just like everything we have," JB told him.

Hat said to wake him up, so Lonny slapped him in the face, and then removed the gag from his mouth. The young man was still asleep, so he received a firmer slap from Lonny, who asked him. "What's your name?" as he was slowly coming around.

Lonny slapped him again and the youth said, "T my name is T."

"No, what's your real name," Lonny growled as he started his slapping motion.

"Tevannel, my name is Tevannel"

"Tevannel, that's not gangster; that's not gangster. You walk up on us, shooting at us and your name is Tevannel? I'm going to change your name to Cujo Wallyo 'because you look like a damn dog!"

Lonny removed all the gold Tevannel was wearing and reached down and took off his Jordan's. Hat told JB to give him more truth serum. Once Cujo went back under, Hat showed him a picture of the crime scene.

"Record this," he told Garries as Maurisee placed one of the mini cameras in the room.

Hat pointed to one of the pictures. "Tell me about this; were you there?"

"Yup" Cujo replied.

"What happened?"

"Leon had us go to the apartment because he'd placed some money under a floor tile and needed it. When we arrived at the apartment, we found Leon's girl in bed with a guy, so we called Leon. We made them put their clothes on – girly was fine! Lover boy took offense to me staring, so we had to dot his eye a couple of times. When Leon arrived, he told us to beat them, so we did that shit. Then Leon jumped on Lover boy and said, "He loved you more than me! He always talked about you and your mother, he drove my mother away, ole' fool even had a will, He left you everything! Now you take my girl and I financed her college, and you sleep with her, my girl!"

Then Leon took out his gun and shot the girl, BOOM, and then he punched Lover boy over and over and finally shot him. He was brave; he died with his signs up."

Tevannel looked at Hat and around the room, and then he fell asleep. Hat looked at everyone and warned them, looking into Marty's eyes "If I die tomorrow, use that information." "Listen fellas, the same DNA required to defend this nation is the same DNA required to avenge my cousin's murder, do not follow me off the cliff, do not follow me down a rabbit hole with no exit, get out of my field of gravity. AB, you were right all along and now everyone knows the truth. My actions compromised our integrity, and I stepped on what we swore to defend." Marty interrupts and ask, "What about your wife and son."

With Rick quietly sobbing, Hat touches the shoulder of Rick and Marty says, "Before my wife and son, you guys were and will always be my brothers, Marty send everyone home." Hat then walked out. Normally, quiet Barnett says "I know I know I know, this is different, but our squad leader would never leave us on the battlefield. We knew the commitment when we took our oath, and we know the commitment now! I am not leaving Hat on this battlefield; I AM ALL IN." Marty waited for Barnett to finish then said, "We have about 12 hours to get everything ready."

Benjamin and I drove Utah and Hat to the hotel, Royster who driven Lynnette and Karen to the hotel pulled guard duty outside the hotel, while Benjamin and I returned to the fortress. Lynnette greeted Hat at the door and told him to take a shower, so Hat placed his gun, keys, wallet, and condoms on the nightstand. Lynnette waited for Hat to get into the shower then using a hair pin poked holes in his condoms.

While Hat was in the shower, Lynnette heard a knock on the door. Utah was standing at the door with a towel wrapped around his midsection, so Lynnette got a good look at Utah's legs. He asked to speak with Hat, but Lynnette answered, "He's in the shower; what do you want him for?"

Utah, never lost for words, can't summon the words. Lynnette assessed the situation, and left Utah at the door while she retrieved Hat's condoms from the nightstand, handing them all to Utah. Utah smiled and walked away.

Back in Utah's room, Karen confessed she'd never been with a white boy before, while kissing Utah's arms. Utah told her it was Ok; "I've never been with a white boy either!" Then Utah reached over to the radio while Karen kissed his legs and turned the radio to a country station. Karen, now on top of Utah, reached over to the radio and slammed the radio onto the floor. Back in Hat's room, kissing his wife Hat reached over to the nightstand to retrieve a condom, and Lynnette told him, "Husband, we don't need to use those anymore, give me another baby, baby."

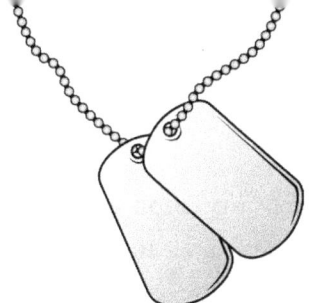

CHAPTER TWELVE

Back at the fortress, Marty, Maurisee, JB and Rick continued to review the pictures recorded by the fly-on-the-wall. I told Lonny today had been a crazy day.

Additional analysis of the recording by the fly-on-the-wall revealed Leon had eight toilets installed, three acid bends designed to destroy hard drives and zip drives for the two laptops, a pool table and four HDTV's hooked up to Play station, XBOX, Wii, and the Playboy channel.

There were two rooms with gigantic beds in them, and another room contained a large, black leather couch where three women in bathing suits were making the crack. There were cameras in each room, and four men on rotating shift inside. Due to the motorcycle boys' indication and warning system, the men in the room would have plenty of time to destroy the drugs and laptops if necessary.

Marty commented, "Leon must have gone to the drug czar school of Nino Brown."

After reviewing the fly recording, Marty, JB, AB, Benjamin, Joey, Maurisee, and Rick formulated a game plan to take down Leon and his operation. Lonny filled Marty in on what had happened at the Police Station. Marty decided not to release Tevannel until the morning.

At 4.00 am, Marty called me, Cousins, Garries, Barnett, Benjamin, AB, Rick, JB, Lonny, Joey and Maurisee together and told us to join hands and bow our heads. Then Marty took a moment before he said "Heavenly Father please forgive me for not praying and talking with you daily. When things go wrong, I should run to you not run away from you! Only you can solve my problems, I was wrong when I stopped talking to you. You are an awesome GOD and I need you in my life always Father GOD, this is real talk GOD" wait Lonny said "can you say real talk to GOD" Garries replied "real talk means you are talking from the heart, GOD just wants to have a relationship with you, talk to him, just talk to him, the scriptures says where two or three are gather together in his name, he is in the mist so talk, GOD is here. " Marty continues "I ask that you please forgive our sins and forgive us for what we are about to do. Thank you for all those years you saved our lives, thank you for your mercy and grace Jesus, let us all say AMEN." After our prayer Marty tap-coded Royster to return to the fortress, and then Barnett drove AB and Benjamin as close as he could to the projects without being seen. AB managed to sneak into the building adjacent to the protected door, and Benjamin was able to sneak into the adjacent building to the left. Both men had their SR25 sniper rifles with them and had changed into a camouflage pattern jumpsuit for the roof. Leon was smart, but he'd never thought to protect the roof. Both men also lugged with them a jamming device; designed to jam all cell phone signals. Both men would remain stationary for seven hours, focusing and shooting when directed. The battle would be run from a Suburban TOC stationed across the street from the projects.

At 9.00 am, Lonny released Tevannel and told him not to go back on the block. Lonny then cut off Tevannel's pants and told him "you wanna wear your pants off your ass showing people your nasty underwear, you really don't need any pants on, walk your sorry ass back just like that." Tevannel looked stupid, the wannabe killer was walking down the street in his underwear with no shoes on his feet. Barnett left at 9.00 am, checking the junk yard for tires, while Maurisee, Garries, and J.B. replaced the batteries in every piece of equipment.

I took Royster to the train station. Royster would take his previous position on the train track beam, but for now he would stand on the train platform with the crowd control device. Rick carefully loaded the explosives into a Suburban. As I drove back to the fortress, I was thinking how we were getting ready to execute something that the insurgents used on us in Iraq.

At 11am, with the streets crowded with gang bangers in anticipation of a battle, Rick set off three car explosions. Each car bomb explosion was rigged for detonation one minute and 30 seconds apart. The three cars Rick chose were on the opposite corners of a square block. After the last car explosion, Barnett placed and set tires on Fire, giving Royster enough time to move from his position on the platform to the train track beam.

Lonny's job was to track up the stairs of the 16-story building where the drugs were being made and place the cell phone jamming

device on the roof. His placement of the last cell phone jammer created the triangulation needed to ensure no one could use their cell phones or landlines. He then returned down to the third floor and took up a position looking outside the third-floor window with instructions to shoot anyone trying to enter the building. My job (Frost) was to cover our six by going to the roof of the adjacent building, just in case someone thought about the roof. What Lonny and Hat didn't know was that Marty knew Hat's and Lonny's characters so on the first night they'd received their biometric 9mm pistols, they were switched by AB.

After the explosions, the gang bangers were running all over the place, but a core group stayed close to the street in front of the main building where the drugs were.

Royster, some 25 feet from the ground, was told to use the crowd control device Maurisee had given him in a sweeping motion against a gathered crowd. The device emits an ear-vibrating signal that causes individuals to place their hands on their ears. Some of the gang bangers had their pistols in their hands, so when Royster pointed and turned on the device, two gang bangers shot themselves trying to place their hands on their ears.

I jumped out of the Suburban with my modified taser/stun gun with ear plugs and shot/tasered/stunned every gang banger standing or kneeling. Then I placed ribbed bans on their wrist, in a police officer cuffed manner, and took their weapons. Several gang bangers returning from looking at the car explosions Fired on me, but the distance was too great as they were running towards me shooting Wild West-style.

Benjamin posted on the roof was able to pick them off easily. Royster, now on the ground, helped me rib band handcuff those individuals who were struck by Benjamin, who was using rubber bullets.

Lonny, wanting to get into the action, attempted to Fire his 9mm from the window in the hallway of the third floor, but his 9mm misfired and Lonny realized he had the wrong weapon. AB had been watching Lonny and used his earpiece to provide that information to Maurisee.

Joey, meanwhile, stayed in the location that housed Mannatorres. The people in the main building were coming out of their apartments because they couldn't use their cell phones, and they saw all the smoke and young gang bangers on the ground.

Maurisee told Lonny to maintain his position as people came outside. When the door of the drug location opened, Leon and all his men walked out, and from the street, through the smoky tires came Hat and Utah. It was 12 noon, and Leon showed just like Mannatorres had said. Rick came out of the Suburban to join Hat and Utah as they walked towards the people now protecting Leon. Leaving his position, Lonny ran downstairs to join Rick, Hat, and Utah, and echoed the same

words he'd said in Iraq, "You people know who are bad in your community, yet you do nothing about it!"

An old black lady answered and said, "Who asked you to come here? Who asked you to protect

us? We were fine before you came here. Go home! GO AWAY!" People in the crowd echoed what she said.

Leon was laughing, as Hat looked on with contempt. Royster ran up to the crowd with the crowd control device but decided not to use it when he saw the elderly adults, as JB told him to return to the Suburban.

Now a larger crowd gathered, and Hat spoke as he pulled out his 9mm, "The man you seek to protect killed my brother. So, you all have a simple choice: join us both in Hell or stay on earth a little while longer!"

Leon, seeing the gun Hat pulled out, pulled out his own gun and raised it in the air, shouting, "I have a gun too!"

AB shot the gun out of Leon's hand,

and the crowd ran off, leaving Hat and Leon. Even Leon's men backtracked about 15 feet before pulling out their guns while attempting to locate the direction where the shots came from. They did not have to look long as AB and Benjamin took aim and shot all their weapons out of their hands.

"You knew Robert was your brother!" Hat shouted.

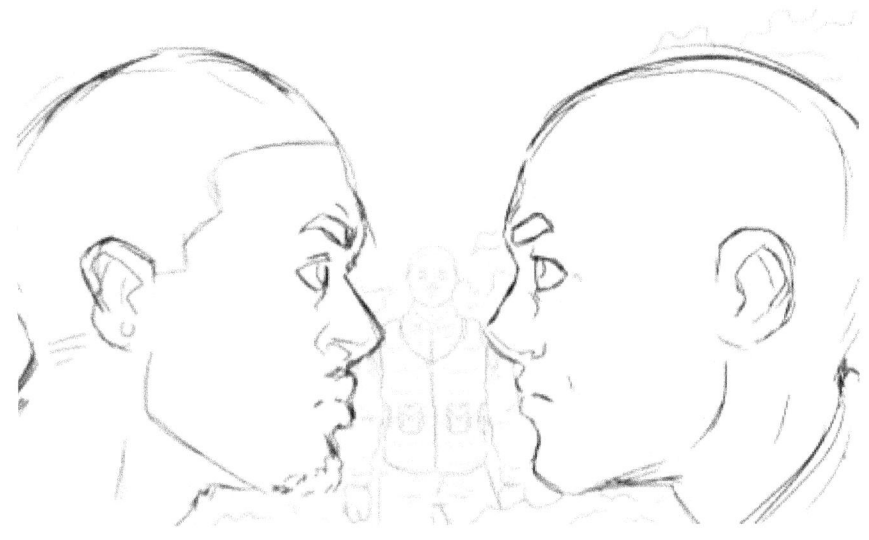

"HALF BROTHER" Leon returned.
"He was your damn Brother, Leon" Hat fired back.
"Whatever he was or wasn't, he is no more" Leon said.
"You also killed your own uncle!" Hat roared.
"He was a weak alcoholic" Leon scoffed.

Marty used his earpiece and told Royster to return to the scene and taser Lonny, who was standing 15 feet to Leon's left, looking for Leon's gun to use against him. Royster walked up and stood next to Lonny and whispered, "Sorry," then tasered him, dragging him back to the Suburban.

Maurisee, Cousins, Garries, and JB all in the Suburban TOC were distracted by the events outside and never noticed a blip on the computer screen. The location of the blip was very close to their location. Finally, JB noticed  the blip on the screen and touched the computer screen to verify the location of everyone. As he touched the screen the names popped up, along with the individuals' locations. JB frowned and told Maurisee "this blip on the screen has a blue force tracking device but it is not ours."

In the heat of the moment, Maurisee decided to inform Utah, and told him, "Go to the building on the right, and head up the stairs after the unidentified person."

The unidentified person was climbing quickly up 16 flights of stairs. Looking at his computer intensely, Maurisee could see that the unidentified climber had already made it up to the 15th floor. In a hurried voice he warned AB that an unidentified individual would shortly invade his position.

Too late, AB and the mystery person were suddenly struggling on top of the roof. Marty asked Benjamin if he could get a shot of the individual and identify him. JB suddenly recalled giving the experimental blue force tracking device injection to a group of newly graduated

snipers, who volunteered to receive the injection after they graduated from sniper school at Fort Benning Georgia, two years ago.

"You think you're not going to pay for all you did?" growled Hat.

"What about you, Soldier boy? You're gonna pay," Leon sneered.

"So, you're what a gangster looks like. You ain't shit!" Hat professed.

"Yeah and you're not in Iraq; you're inside my wire, HERO!" Leon proclaimed.

"We brought Iraq to you," was all Hat said.

Benjamin answered Maurisee and said, "I can see only half the fight. My view is blocked by an object on the roof; I have a better view of the street."

"It's OK," Marty told him, "AB should be able to handle himself." AB was still struggling with the mystery man and Utah was struggling with climbing the stairs with those prosthetic legs.

Hat lost his thought and said, "What about the little girl?"

"What little girl?" replied Leon?

"The little girl you hid in the wall!" Hat yelled, and then he struck first, punching Leon in the face and knocking him down, then jumping on top of him. Leon's men cheered Leon to fight hard. Rick taking advantage of their attention on the fight sneaked around towards the room with the drugs in them and told the girls to put on some clothes and leave quietly. While Hat was fighting, Rick took all the zip drives from the two laptops and started destroying the drugs by flushing them down the toilet.

From his vantage point on the roof, Benjamin saw the Police and Fire Department heading towards the projects. Maurisee asked Benjamin to double check the fortress to ensure no one was present. "The coast is clear," Benjamin confirmed so Maurisee pushed a button and blew up the fortress. Within a few seconds, the Fire Department and Police diverted towards the explosion at the fortress.

AB lost his battle with the unidentified person on the roof, who pulled out his SR25 and prepared to shoot Hat, who was still fighting

Leon. Benjamin told Maurisee to give Royster some smoke grenades and throw them at Hat and Leon. "If he's a sniper, he won't shoot what he can't see."

Royster ran and popped smoke near the fight. The rooftop sniper took aim but was unable to identify his target but decided to pull

the trigger anyway. The smoke screen gave Utah enough time to reach the roof and see the unidentified man take aim in the smoke and pull the trigger. Utah jumped on the mystery man and placed him in a grappling hold,

and then Utah pulled his shoulder out of the socket, and then lifted the mystery man above his head and throw him off the roof. Then he knelt and administered first aid to AB.

Mannatorres on the move, is followed by Joey.

The blind shot caught Hat in his butt, and Leon was using Hat's loss of strength to his advantage and was beating Hat.

Another blip on the screen popped up, but Maurisee and JB had already left the Suburban as Cousins and Garries eyes were glued on the fight outside. As the smoke cleared, Utah was helping AB down the stairs and out of the building. Marty and I walked up to the fight to see Leon holding Hat in a headlock. Leon's boys were yelling and aiding Leon by stomping on Hat. Marty calmly walked up and shot one of the guy's stomping Hat.

"This is a fair fight" Marty said, then shot the other guy stomping Hat. Royster cuffed the two men Marty shot with rubber bullets.

Now everything was frozen, as the third guy backed away and Leon released his choke hold on Hat and stood up. Lonny stumbled out of the Suburban in time to hear Marty say to Hat, "Lonny was right, we should have just shot this piece of shit and been done with it. If these people want to live like this, why should we stop them?"

Marty, looking at Leon, said "Here! You want to kill my best friend?" then tossed his gun at Leon.

Marty growled "This man here pointing to Hat is a no limit soldier." Those pieces of shit; pointing to the men he shot, are just that, shit!" Pointing now to Hat, he continued, "He's your cousin! Pointing to Leon Marty said "Man, you have killed your own people, your own KIN well, finish the job and shoot him!"

Leon looked at Hat, then at Marty. Leon pointed the pistol towards Hat and said, "I'm going to shoot him, then I'm going to shoot him!" pointing to Rick, who had just come out of the building after destroying the drugs, "and then I'm going to shoot you for being such a damn dumb ass fool!"

Leon backed up, took aim at Hat's head and squeezed the trigger. The gun was out of rubber bullets, but a shot was heard as Leon's body was blown back, with blood coming from his chest. As everyone turned towards the shot that rang out, the blip no one noticed was Lynnette, who was standing there hold-  ing the gun aimed at Leon, with Karen standing behind her. Marty turned and said, "No, you're a damn dumb ass!" to Leon's fallen body, shaking on the floor.

Mannatorres is within four city blocks and Joey decides to use his cell phone to call Maurisee about Manatorres but, he is within range of the cell phone jamming devices and can't reach Maurisee or anyone else.

 Lonny took the gun from Lynnette's hand and walked slowly over to Leon and just as he did in Iraq, shot the wounded, shaking body of Leon. As Lynnette attempted to help Hat up, sirens were racing down the street and Mannatorres, the MPs from Fort Hamilton and the New York City Police with Officer Nelson were all in tow.

Mannatorres pointed to all of us, telling the MP's and Police we'd murdered people. Rick said angrily, "What about these drugs you bring into this place? What about your operation? What about how you muscle everyone in this neighborhood and take our money?"

Mannatorres said, "If that was true and it isn't, your papa must be pretty rich."

"What about these?" Rick asked, holding up the zip drives, "I bet they have information about you on them."

Mannatorres smiled. "They will never see a day in court, and neither will I, you dumb stupid Jewish pig. You're nothing. You're worse than a pig to me. You're crazy; no one can touch a Mannatorres!" expressing himself with braggadocio bravado.

Rick looked deflated, defeated, and rejected and looked down, so he missed seeing Mannatorres' body fly backwards. Benjamin, still on the roof, loaded a real bullet in his rifle and aired out Mannatorres.

When Rick looked up, Mannatorres was sprawling on the ground. The Police and MPs reacted and ran towards the direction of the shot, while Mannatorres' body guards stayed with Mannatorres. Benjamin then parachuted off the roof. Frost seeing Benjamin parachuting off the building decided to aussie rappel down to the ground.

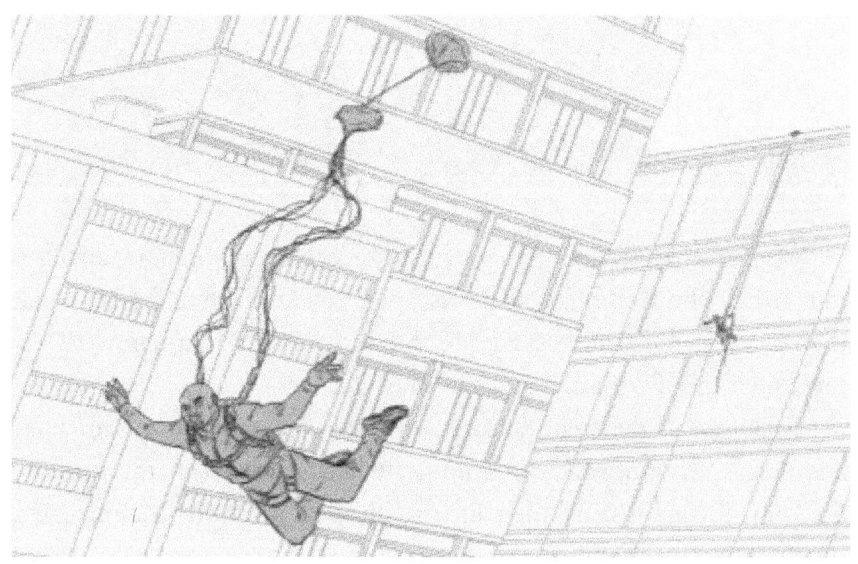

Hat, still bleeding from his gunshot wound, called for Marty and Rick and with Lynnette holding him up said, "Get them all out of here. I am not running; it's time to go."

"No!" said Rick and Marty simultaneously.

Looking at Marty, Hat just said, "The needs of the many…"

"This isn't Star Trek" Marty cut in.

"Donald Gordon" Hat replied.

Marty told everyone to get in the Suburban's. Utah placed AB inside the Suburban, and then joined Karen, who was standing next to Lynnette. The Suburban's drove off, leaving Hat, Lynnette, Karen, Utah, and Rick behind.

Officer Nelson returned from his short jog towards the building where the shot came from and placed Rick, Hat, and Utah under arrest and into his patrol car. As the MP's returned, realizing there was no need to run towards the shots fired, Officer Nelson told them, "Report to the precinct to pick these Soldiers up. I'll release them once we book them" and the MP's drove off towards the Police Station.

Joey followed Maurisee who picked up Benjamin and Frost at the rally point, which was the blown-up fortress. The Fire Department was just putting the finishing touches on the fire Maurisee started. Marty doubled back and picked up Lynnette and Karen. Officer Nelson surprisingly drove past the precinct where the MP's were standing on the steps, and took Hat, Rick, and Utah to the Brooklyn Bridge.

There, he stopped and helped everyone get out of the patrol car. As he took off their handcuffs, he said to Hat, "You're a failed leader. You are supposed to rise above this shit – you're better than this. You lost your core values and endangered your men for your own purpose. Your job is to support and defend this great nation; you're a Soldier but you abused your power with principle when you went after Leon. Core value number four, selfless service put the welfare of the nation, the Army, and your subordinates before your own. The next time you decide you want to fight crime and kill drug dealers, get out of the military, and join the New York Police Department, we wrote the book."

Officer Nelson then reached into his pocket and threw his All American 82nd Airborne patch at Hat, then drove off.

Marty got directions from Maurisee and picked up Hat, Rick and Utah.

The next day, the old lady that was mouthing off when the crowd gathered to protect Leon was outside trying to make a call on her cell phone. She still couldn't hear anything as Maurisee had forgotten to turn off the button that activated the cell phone jammer.

The end

Heavenly Father,

I want to thank you for your fresh grace and mercy. Heavenly Father, I ask that you forgive me of my sins. I believe in my heart that JESUS CHRIST is my LORD and Savior. I fail and fall short every day, Lord Jesus, and I am not worthy of the blessings you have bestowed upon me and my family.

I worship you, Lord, because you love me more than I love myself. I worship you, Lord because you are the Alpha and the Omega. I worship and praise you, Lord, because you are God Almighty the everlasting, the Creator of all things. I thank you for the lessons of yesterday, today, and tomorrow, knowing nothing is promised to me but your everlasting love. I want to thank you for our relationship and our daily conversations.

I ask that you allow me to hear your voice louder, and louder. I ask Lord that you send your angels to encamp and protect my entire family. Please send your angel of mercy to oversee my kin folk and friends.

In the mighty name of Jesus, I pray, Amen.

Tabatha,

I cannot see myself without you; when I look in the mirror, I see you. You know the man that I am. I become half the man I am when we are apart; I need you to make me whole. Puddin' my beauty, you are my soul mate, created just for me. You are not a background singer but a leading lady who stands beside me no matter what the challenges are.

You don't just love me in every way possible, you inspire me and help me mold my dreams, and my dreams became your dreams, and those dreams turned into reality. I understand those career sacrifices you have made on behalf of our family and my career. I never would have reached Command Sergeant Major without you, and I simply adore you.

You put your career on hold as we moved from country to country and state to state every 24 months. Man, when God blesses you, he gives you more than you can imagine, and I am so happy he gave me you. Thank you for being my wife, lover, queen, teammate, Mother of my children, and my best friend. I love you.

POSTSCRIPT

When GOD called you home on April 7th, 2014, I was shocked and broken. It has taken me six years to place illustrations into this book. Today, I can say, thank you Jesus for the 24 years you allowed me to be with Tabatha. You are an awesome GOD!

My Children,

Zachery (Big Country)
Spencer (Big Daddy)
Karina (Baby girl)
James (The General AKA Master AKA Five Star)

I am very proud of you all and I love you. There is greatness in all of you. All of you are bound in great power. Your single and simple principal objective is to be better than me. From your education level to your finances, you must do better.

It is important you move our generation forward to the next level. Never forget the environment I was raised in. Your Grandmother always told me to be better than her. You all are now charged with the same task. From the projects to mansions, THAT IS THE GOAL, STAY FOCUSED.

Please note that I have stumbled and failed from time to time, and you will too, but keep on going. Remember you have the blood of your ancestors. Be your own leader, dance to your own drum, and never worry about what people say about you. People talked about you when you were born and people will talk about you when you die, but the only thing that matters is I LOVE YOU! Expect nothing and trust GOD always!

Ma,

If I begin to thank you for everything you have done for me, I would definitely have to write a book in your honor. You worked two jobs for 17 years to provide for Towanna and I. I cannot thank you enough for my great childhood and the many lessons I have learned through your integrity, dignity, and work ethic.

I am reminded of my first day in the first grade. The teacher asked each child what they wanted to be. Some said President, School teacher, Fire Fighter, Doctor, Policeman, and Nurse. I said, "I want to be a mountain climber." Some kids snickered, some laughed outwardly. Upset with my classmates' response, I waited until you arrived home to tell you what happened. You told me, "Boy, people talked about you when you were born and people will talk about you when you die, the only thing that matters is I LOVE YOU. When I climbed Mount Fuji (the tallest mountain in Japan) in 1997, I thought of you. I am not sure if my first-grade classmates achieved their desired profession, but this I do know, it took me 6 hours and 38 minutes to reach that summit, I love you Ma.

POSTSCRIPT

When the GOOD LORD called you home on May 4th, 2013, I was lost. Seeing you in the hospital for five months, I thought I was prepared mentally for your physical absence. I was wrong. After seven years, I still hear your voice and I still see your face and I always will.

James III,

Sir, when you departed out of my life at the age of seven, I was transmogrified and forced to mature quickly. I do not fault you for your actions. I want you to know that I have never smoked dope or taken drugs. I've watched over my sister, and never disrespected my mother. I hold no malice towards you, and I forgive you. You're my father. I pray for you daily and I love you.

Charles,

Dad, I think you are the greatest man on the planet. Decades ago, you came to this country from Haiti and worked your butt off driving a cab in Manhattan for over 20 years. Those stories you told me about driving your Taxicab on 9/11 remain with me. Courage runs in our family.

I want to thank you for being an outstanding grandfather and for marrying my mother. You are a quiet gentle giant, and I believe my mother is still upset with me for explaining and introducing sports to you (smile). I love you Grandpa Charles.

Towanna,

Baby Sister, you have several bachelor's and master's Degrees, yet your desire in life was to return to our neighborhood as a teacher and now counselor. I am proud of you beyond words.

Remember that time on Christmas Eve? Ma put us to bed and when she closed the door, we went to the window looking for Santa Claus.

We lived on the ninth floor; I was eight and you were five. You claimed you saw Santa Claus on the rooftop of a building, and you convinced me so well that I saw him also. I was never a bright kid, because a little girl with big bifocals should not be trusted to spot anything a mile away. When I asked you how Santa gets into apartments when we did not have a chimney you said, "Stupid, Santa has a magic key." Oh, was my response, then I thought, "I hope Santa doesn't lose that key because if he does, we are going to get robbed."

The skin on my body would like to thank you for insisting I get home before curfew. You were smart then and you are beautiful and smart now! I love you- my Baby Sister.

Edward,

B-Law, I want to thank you for marrying my little Sister. I am glad you came into her life at the right time for the right reason. You overwhelmed her with your loving kindness, and I could not pick a better man for my sister.

Thank you for treating Towanna like your Queen. I appreciate the fact that no matter where I am in the world, on Veterans' Day and Memorial Day, you find me and thank me for my service to our great nation. I love you like a brother.

Charmaine,

Sister-in-Law, I love you so much! When you found out one of your nephews was asked by his first-grade teacher what he wanted to be when he grows up and he replied, "A horse," you said seriously, "Well damn it Son, you be the best damn horse you can be!" You are a trip and a joy to be around. I must confess when I am out of the country and feeling a little sorry for myself, you send timely short emails telling me you love me. Thank you, I love you back.

Clayton,

B-LAW, I want to thank you for being my sounding board. I am built to deal with a great amount of stress. Showing weakness to those I must protect and provide for is not normal for me. Thank you for listening. One love B-Law'

KIN FOLK,

I would like to acknowledge the following families: the Thorpe's, Garries, Cousins, Jones, Hawkins, Soles, Jackson, Craigs, Josey's, Ellington's, Jones'. Charles R. Jones- Grandpa, Joseph Cousins- Grandpa, Ellen Cousins- Grandmother, Mary Jones- grandmother and, Elverta Woodson-grandmother.(Thank you for watching over me from heaven, tell Grandpa Charlie and Grandpa Joseph I said hello…. two of hearts!), nieces Taniqua Hunter, August Lett, Chelsea, nephews Stephon and Trevon. Aunts= Agnes Simmons, Lucille Vick, Christine Harrell, Mary Ford(uncle Mike), Diane Morris, Pat Taylor(uncle John), Joanne McDowell, Ellen Marie Oliver, Pam Cousins, Belinda(uncle Gregory),Noreen(uncle Ezell) Chareen(uncle Viven), Michelle, Fabeola (Special Aunt-Aunt Esther)Uncles-Spencer & Chuck Jr. (Special Cousin-Candice Craig) Cousins-Georgia Josey, Carl Craig, Kenneth Craig, Troy (Karima) Craig, Tiquoya St. Louis, Tasha, Terrence, Terrell, and Tylon Stockton, Gerald Blayne Jr, Malcolm, Ahasuerus McDowell, John and Andre Taylor, Sade, Michael, Sequan, Tiffany Morris, William,

Faye, Marcel, Joseph, and Michelle, Eric Taylor, Matthew and Maurice Harrell, Chanell, Donnell, Kim, Richard, Quason, Spencer Jr. Robert "Fun" (Germerish) Wise, Lisa Wise, Fannie, Preston, Pervis, Arthur, Ali Khan, Arthur Jr, Allika, Ashia, Akilah, Terrence, Chris, Joel, Gaila, Candice, Adarine, Dayshawn Jones, David (Gigi) Thorpe, Aunt Elaine, Regina(Tony) Johnson, Elaine Thorpe, Kim Josey, Clifford "Skeeter" Josey, Anthony and Terrell Josey, Chuck Jr. Kyle, Kerri, Vanessa, Gigi, Velicia, Gregory "Tyga", Lamar, Barbara, Pauline (Peaches)-Tuck family. Gatling family. Arthur Ellington "Pops", brothers Maurice and Mike-(Family)Maggie Charles (family). Our ancestry-The Blackfoot Tribe. I want to acknowledge Archbishop Roy E. Brown who imposed the word of God to my mother and involved us in the many activities of the church. Respectfully- Pastor Loncke, Bishop Shorts, & Pastor Johnny. To my church family at First Corinthian Baptist Church and my Pastor Rev. C.E. Salter and Mrs. Salter, thank you for the word of God. To Pastor Jomo and Charmaine Cousins of Love First Christian Center and my church family, thank you and I love you all! I want to formerly request that Pastor Jomo Cousins preside over my funeral. I am positive my Heavenly Father will direct your words, for you are truly a man of GOD.

NEXT OF KIN,

Shantel(Norman)Childs-Williams, Ms. Janice, Charisse & Charlie Kearse, Ira ,Kellye, David Stennett, Nadine, Deirdre "Tink", Audra Hollins, Lindsay Glover, Shirl, Marcella & Ms. Dot, Makeda, Valerie "Cookie", Reggie, DeeDee Falls & family, Ann, Loretta, Cathy Covington(babysitter), Fay, Debbie, Cathy, Irving, Leon Singleton (Ms. Bessie) Family, Willie (Mickey) Randolph & Family, Nancy Everett and family (Phillip, James (Birdie), Reuben (Blue), Leon (Chilly) Everett, Artie (Rosie) Thompson, Desi, Vanessa Cummings, Joan Smalls, Giselle & Albert (Bay Bay) Jones, Debbie, Chanel & Gail Williams, Tiny & Danny boy Cooke, Anthony Henderson, Pop, Bobby and Tyrone Lewis, Lynette Davis, Brenda & Squirt Lasane, Pam Hill, The Shepherd Family, Tracy, Lamont, PeeWee, Robert Lee Family, Margie Thompson, Melissa (Mel Zee), Gretchen Jefferson, Michelle Melendez, Dina Alston & family, Nancy & Andrew Washington, Nikki, Monifa Morgan, Stan & Mary Perkins, Terrie & Ken Dickerson, Yvette Padilla, Paulette & Wendy Brothers, David Collins & family, Judi Wilder and family, Cynthia, Eric, and Eddie Billups, Allan and Tony Thompson, Al Gleaton, Gus Alexander, Lora Anderson, Cecile Beaumont, Edward Evans, Carol Black, Ike Parker, Alisa Thomas, Eleanor Lowry Debra Cline, Derek Stalling, Patrick Jennings, George Perry, Dorothy Croom, Lohntraya Miller, Artemio Camacho, Robert Lewis, Richard Super, Jamal Hashim, Jeff McClaskey, Marilyn Brendalen, Doug Whetstone, Raphael Haddad, Karim Lahdu, Wajih Harroum, Majed Derabuni, Gulsen Beyatli, Hashim Mohamed, Michael Khouri, Mario Elia, Riyad Eldin, Jamal Hashim, Chucky Daniels, Benjamin Hart, Al Pollard, Stewart Dirton, Tommy Ingram, Jerome Hopkins, Robin Christian, Michael Dasovich, Michael and Maria Dean, Derek and Maria Coleman; Jeffrey Ross, Darryl and Tracy Hudson, William West, Karen Douglass, Ellis Warren, Alan Flood, Shawn Hill, Richard C. and Ellen Hoehne, Robert Irvin (thank you), Jacqueline (my mentor) Moate, Queen (my Sista) Risher, Jesse and Toni Pisciotta, Jeanette Barreto, Rudy and Sara Moore, Marlon Rudrow, Martin Middleton, Leroy Nelson, Benjamin Saint James, Luis & Sharon Burgos, Angel Oquendo, Faith Quick, Quintin Royster, Ron Ward, Trancell ,Shenita Scott, Tanya Sullivan, Carlotta Kennedy, Joseph Driscoll, Hammie Session III, Daniel Munchbach, Pastor Marshall & First Lady Selina

Lewis, Arlie Rogers, Dr. Daniel Alcide, Al Dileonardo, Howard Baum (thank you), Kent Johnson (much thanks), Brad Kane (I mentored you, now your mentoring me), Andrea Barber, Sabrina Johnson, Kevin Knuth, Douglas Nelson, Alaina Barnes, Dean Bianco, Frank Caraballo, James Davis, Bryan Farley, Ted Harrisberger (aka the machine), Kara Perkins, Laura Shaner, John Stroncheck, Marna Tracy, and Kevin Garcia.

To the many Military personnel, I have worked with,

Thank you for lifting me on your shoulders.

My teachers,

I want to acknowledge the many teachers and professors I've encountered: Public School 284, Gershwin Junior High School, Canarsie High School, City College of Chicago, Campbell University, Maryland University, Fayetteville State University, Chaminade University and Argosy University.

Finally, to Dr. Gail Demery and Associates,

Gail, you gave me an opportunity to express myself by putting thought to paper. After looking for the right individual to guide me through the literary process, I found you. I should have known the Almighty would only allow a God-fearing woman permission to crack open the door for me.

All I ever needed was an opportunity. I am humbled and grateful beyond words. You are a master motivator and possess all the tools of a polite, demanding, task master. I have had numerous titles associated with my name; through your hard work I can now add "author". Thank you!

If I omitted any family or friend's name, I apologize. It was not intended. Contact any Kin Folk and I will include you in my next book.

أطلب من الله أن ينعم بخيراته لكل من يقرأ هذا الكتاب

ONE FAMILY-ONE LOVE

www.ingramcontent.com/pod-product-compliance
Lightning Source LLC
Chambersburg PA
CBHW071924290426
44110CB00013B/1464